WITHDRAWN

~~GARY LIBRARY~~
~~VERMONT COLLEGE~~
~~36 COLLEGE STREET~~
~~MONTPELIER, VT 05602~~

D1613530

Human Carrying Capacity of the Brazilian Rainforest

HUMAN CARRYING CAPACITY of the BRAZILIAN RAINFOREST

Philip M. Fearnside

New York   Columbia University Press   *1986*

Library of Congress Cataloging in Publication Data

Fearnside, Philip M. (Philip Martin)
  Human carrying capacity of the Brazilian rainforest.

  Bibliography: p.
  Includes index.
  1. Environmental policy—Amazon River Region.
2. Amazon River Region—Population density. 3. Land capability for agriculture—Amazon River Region.
4. Rain forest ecology—Amazon River Region.
5. Rodovia Transamazônica (Brazil) I. Title.
HC188.A5F42   1986   330.981'1063   85-6643
ISBN 0-231-06104-8

Columbia University Press
New York   Guildford, Surrey
Copyright © 1986 Columbia University Press
All rights reserved

Printed in the United States of America

Clothbound editions of Columbia University Press books are Smyth-sewn and printed on permanent and durable acid-free paper.

Para o povo da Amazônia

Contents

	Preface	*xiii*
	Acknowledgments	*xv*
1.	**Development Rush in the Amazon Rainforest**	1
	The Shrinking Rainforest Areas	2
	Brazil's Colonization of the Amazon	15
2.	**The Tropical Rainforest As an Ecosystem**	35
	Ecological Features	35
	Environmental Concerns of Rainforest Use	42
	Effects of Agricultural Operations on Soil	53
3.	**Population Growth and Carrying Capacity**	61
	Human Population in Ecological Communities	61
	Population Growth Patterns	63
	Factors Limiting Population Growth	68
	Definitions of Carrying Capacity	70
	Instantaneous Carrying Capacities	70
	Sustainable Carrying Capacities	73
	An Operational Definition of Carrying Capacity	79
	Assumptions in Carrying Capacity Estimates	80
	Models for Estimating Carrying Capacity	86
4.	**Modeling the Agroecosystem of the Transamazon Highway Colonists**	93
	Location and Description of the Study Area	93
	Comparability with Other Tropical Areas	96
	Methods	100
	Modeling Crop Yields	103
	Land-Use Allocation Sector	110
	Product Allocation Sector	113
	Population Sector	116

5. **A Stochastic Model for Human Carrying Capacity on the Transamazon Highway** 121
   Features of the KPROG2 Carrying Capacity Model 121
   Modeling Methods 126
   Model Structure 126
   Simulation Results 130
   Conclusions on Carrying Capacity 144
6. **Choosing Development Strategies for Rainforest Areas** 147
   Defining Development Objectives 147
   Conflicts of Objectives 150
   Colonization Programs and the Fulfillment of Objectives 152
   Carrying Capacity and Development Policy 155

Appendix: Summary of KPROG2 Model Parameters and Equations 159

Notes 227
Glossary of Technical Terms, Foreign Words, and Abbreviations 231
Bibliography 241
Index 281

Illustrations

Figures		Page
1.1.	Map of Brazil's Legal Amazon	2
1.2.	Population Increase in Rondônia	6
1.3.	Agrovila Grande Esperança	21
1.4.	Colonist Family Living in Lot	22
1.5.	Rice Fields in Newly Cleared Land	24
1.6.	Typical View of the Transamazon Highway	25
2.1.	Poor Burn	41
3.1.	Exponential and Logistic Growth Curves	64
3.2.	Operational Definition of Carrying Capacity in Terms of a Gradient of Failure Probabilities	80
3.3.	Hypothetical Effect of Variability on Failure Probabilities and Carrying Capacity	82
4.1.	Map of the Transamazon Highway Intensive Study Area	94
4.2.	Poor Transportation Conditions in Lateral Roads	95
4.3.	Map of Soil pH Under Virgin Forest in the Transamazon Highway Intensive Study Area	96
4.4.	Mean Monthly Rainfalls for Altamira (1931–1976)	99
5.1.	Sectors of KPROG2	127
5.2.	Simplified Causal Loop Diagram of KPROG2	128
5.3.	Causal Loop Diagram of KPROG2	129
5.4.	Flow Chart of KPROG2 with Operations Grouped by Level	131
5.5.	Simulated Calories per Capita	134
5.6.	Proportion of Simulated Lots Below Calories Standard	134
5.7.	Proportion of Simulated Area Cleared	135
5.8.	Proportion of Simulated Lots Below Calories Standard versus Areawide Average Calories per Capita	136
5.9.	Colonist Failure Probabilities versus Population Density	137
5.10.	Simulated Land Use by Colonist Type	142
5.11.	Colonist Type and Failure Probabilities	143
5.12.	Relation of Population Density to Combined Probabilities of Colonist Failure by Colonist Type	144

Tables

1.1.	SUDAM Estimates of Amazon Forest Clearing	3
1.2.	LANDSAT Surveys of Forest Clearing in the Brazilian Amazon	5
1.3.	Parks and Reserves in the Brazilian Amazon	8
1.4.	Estimates of Extent of Pasture Degradation in Amazonia	27
2.1.	Major Ecological Zones of the Brazilian Amazon	36
2.2.	Major Soil Types of Transamazon Highway Colonization Areas	39
2.3.	Possible Macro-Climatic Effects of Amazonian Deforestation	51
2.4.	Summary of Effects of Agricultural Operations on Soil	59
3.1.	Population Doubling Times for Selected Areas	65
3.2.	The Two Methods	89
4.1.	Rice Yield Regression Excluded Category Summary	105
4.2.	Maize Yield Summary	108
4.3.	Data for Calculating Technological Change Factors	109
4.4.	Land-Use Probabilities Based on Colonist Type	111
4.5.	Probabilities of Cash Crop Land Use	112
4.6.	Colonist Diets	115
4.7.	Frequencies of Colonist Types in Original and Newcomer Populations	116
5.1.	KPROG2 Program Operations by Level and Sector	132
A.1.	pH in Initial Soil Quality Generation	161
A.2.	Transition Probabilities for Virgin Soil pH	162
A.3.	Clay for Initial Soil Quality Generation	162
A.4.	Slope for Initial Soil Quality Generation	163
A.5.	Carbon for Initial Soil Quality Generation	163
A.6.	Phosphorus for Initial Soil Quality Generation	163
A.7.	Monthly Rainfalls as Proportions of Period Totals	164
A.8.	Variability in Daily Weather as Proportion of Monthly Totals	165
A.9.	Financing: Frequencies, Amounts, and Terms	166
A.10.	Total Labor Requirements for Agricultural Operations by Month	177

A.11.	Male Labor Requirements for Agricultural Operations by Month	178
A.12.	Labor Requirements for Agricultural Tasks	179
A.13.	Fixed Cash Costs for Agricultural Operations	183
A.14.	Virgin Forest Felling-Month Distribution	186
A.15.	Second Growth Cutting and Burning Month Distributions	187
A.16.	Government Fertilizer Recommendations for Pepper	196
A.17.	Government Fertilizer Recommendations for Cacao	197
A.18.	Prices of Fertilizers and Lime in Altamira	198
A.19.	Soil Level Decreases Under Other Uses	200
A.20.	*Phaseolus* Bean Yield Regression Variables Summary	202
A.21.	*Phaseolus* Bean Yield Regression Excluded Condition Summary	202
A.22.	*Vigna* Cowpea Regression Variables and Excluded Conditions Summary	203
A.23.	Outside Labor Type Frequencies	209
A.24.	Outside Labor: Days Spent and Earnings by Labor Type	210
A.25.	Cash Sent and Received from Outside Area	211
A.26.	Spoilage of Stored Products	212
A.27.	Transportation Availability	213
A.27a.	Prices of Products	213
A.28.	Price and Nutrition of Principal Goods Bought with Cash	215
A.29.	Seed Requirements	217
A.30.	Proportions of Free Capital Invested After Satisfying Subsistence Needs	217
A.31.	Proportions of Investment Capital Within Each Category Spent on Purchase of Capital Goods	218
A.32	Initial Population Characteristics (at arrival)	218
A.33.	Depreciation of Capital Goods	219
A.34.	Return on Capital Goods Relative to Manual Labor	219
A.35.	Demographic Information	220
A.36.	Initial Capital and Capital Goods	222
A.37.	Labor Equivalents in Agricultural Work	222
A.38.	Disease Probabilities by Age and Sex	223
A.39.	Monthly Disease Probabilities	223
A.40.	Work Days Lost to Illness	224
A.41.	Age-Specific Fertility for Rural Brazilian Population	224
A.42.	Calorie and Total Protein Requirements and Effects on Mortality	225
A.43.	Probability per Year of Family Emigration	226

Preface

Humans, regardless of their technological level, are an integral part of the ecosystems in which they reside. This book deals with the colonization of Brazil's Amazon rainforest and how human populations fit into the new ecological systems being established there. Estimation of carrying capacity, the density of the human population that can be supported at an acceptable standard of living on a sustainable basis, is vital to any effort to maintain the colonist's living standards and to avoid degradation of the environment.

The book is based on my work over the past eleven years on Brazil's Transamazon Highway, especially data collected during two years of residence (1974–1976) in one of the planned agricultural villages, or agrovilas, built by the Brazilian government during this historic colonization effort. The estimation of human carrying capacity in this area was the focus of my doctoral dissertation in the Division of Biological Sciences of the University of Michigan. I continue to work on problems related to the estimation of human carrying capacity in this area, as well as in other colonization areas in Amazonia, especially in Rondônia.

The present colonization of the Brazilian Amazon is outlined at the outset of the book. Then the ecology of tropical rainforests is summarized as it affects pioneer farming, including the environmental concerns linked with felling the forest for agriculture. Human settlement is put into the context of ecological systems in general, and the various techniques used for human carrying capacity estimation are considered in developing a methodology appropriate for the Transamazon Highway study. The agricultural system of the colonists is then examined for modeling as a part of carrying capacity simulations. A model for estimating carrying capacity is presented, including resource allocation, product allocation, and population sectors, in addition to agricultural production. Probabilistic (stochastic) models are used for evaluating the importance of variability in crop yields on carrying capacity.

It is hoped that the book will be useful for planners facing immediate decisions in areas like the Brazilian Amazon and that it will contribute to developing a sorely needed area of ecological research: an adequate science of carrying capacity.

Acknowledgments

Dr. Paulo E. Vanzolini deserves special thanks for help and encouragement from the beginning of my work in Brazil. I thank the Centro de Pesquisas Agropecuárias do Trópico Úmido of the Empresa Brasileira de Pesquisa Agropecuária for logistical support as well as the chemical analysis of soil samples. The Instituto Nacional de Colonização e Reforma Agrária, the Empresa Brasileira de Assistência Técnica e Extensão Rural, the Museu Paraense Emílio Goeldi, and the Instituto Nacional de Pesquisas da Amazônia have all contributed logistical support during fieldwork. Funds for various parts of the project have come from National Science Foundation dissertation improvement grant GS-422869, a Resources for the Future predoctoral fellowship, two fellowships from the Institute for Environmental Quality, the University of Michigan, and three grants from the Programa do Trópico Úmido of the Conselho Nacional de Desenvolvimento Científico e Tecnológico (including one funded by Projeto POLONOROESTE).

Portions of the text are adapted from papers of mine in previous publications cited in the Bibliography: Fearnside 1979a,b; 1982; 1983a,b,c; 1984a,c; 1985a,b; n.d.(b)(c)(e). Tables and figures that have also appeared in other publications are so identified in table notes and captions.

None of the views expressed are the responsibility of the organizations that have supported the project, nor of the many individuals who have contributed their comments and suggestions. The manuscript has benefitted greatly from J. G. Gunn and W. W. Fearnside's careful reading in the final stages. Thanks are due to all of these people, especially to the colonists in the intensive study area. All errors are solely my responsibility.

CHAPTER ONE

Development Rush in the Amazon Rainforest

THE AMAZON REGION of Brazil, the largest area of tropical rainforest on earth, is presently the site of one of the great discontinuities in human history—the irrevocable replacement of a major biome with human agricultural systems. Rapidly increasing human population throughout Brazil, combined with a strong historical trend toward concentration of wealth and landholdings in settled areas, has created tremendous pressure on people to migrate to less populated regions. At the same time, the opening of highways into rainforest areas has presented the disposessed drought victim and the large corporation alike with the possibility of owning a piece of Amazonia's most sought-after resource: land. In addition to the many human and ecological stresses, this process leads inexorably to a fundamental conflict—the incompatibility of infinite demands with finite resources.

Central to the problem of human populations and their resources is the concept of carrying capacity—the number of people that can be supported indefinitely in an area at a given standard of living without environmental degradation, given appropriate assumptions concerning technology and consumption habits. Exceeding carrying capacity can lead to failure to maintain an acceptable standard of living and to environmental degradation in many forms. Brazil, like other countries with similar problems, can avoid these kinds of consequences, and the human suffering that they entail, through appropriate planning measures. These measures require a better understanding of human carrying capacities: what they are, what factors affect them, and how they can be estimated.

This book is directed at the problems surrounding the estimation of human carrying capacity for a specific rainforest area being colonized in the Brazilian Amazon. Computer simulations are used to investigate factors affecting carrying capacity in a stretch of the Transamazon Highway (BR-230), the road built from east to west across Brazil to stimulate colonization and economic development in the Amazon rainforest (figure 1.1).[1] It is hoped that this study will both contribute to the ability of planners to estimate car-

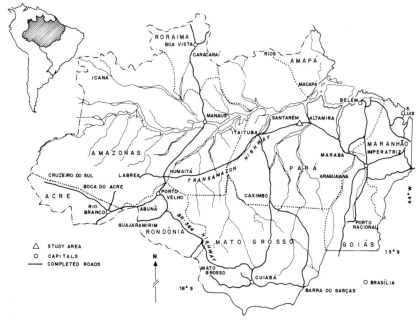

Figure 1.1. Brazil's Legal Amazon, showing existing and proposed highway routes.
SOURCE: Fearnside n.d.(e).

rying capacity and systematically integrate the diverse types of information required for such studies. Both practical results and the theoretical underpinnings of carrying capacity estimation are important since at the scale of current developments in the Amazon, even small amounts of progress can have an impact on the lives of a great many people, both present and future.

The Shrinking Rainforest Areas

THE DECEPTIVENESS OF SIZE

Recent publications reflect widespread concern over the rate at which tropical rainforest, including Brazil's, is disappearing.[2] Nevertheless, the belief persists that the Brazilian rainforest is without end. In 1980, a taxi driver on the Transamazon Highway in Altamira expressed this viewpoint eloquently when he told me that from the air the highway looked like a "mere trail of leafcutter ants," hardly making a dent in the immense forest. More influential people have expressed similar sentiments after making the flight from Brasília to Manaus. The vastness of Amazonia does lead to the impression that it is endless, encouraging people to discount any warnings concerning deforestation. The portion of Brazil legally classified as Amazonia has an area of about five million square kilometers (km^2), about 2.5 million of which are in *terra firme* (high ground) areas where the original vegetation is rainforest (Pandolfo 1978:11, citing FAO data). The rainforest area is approximately one-third the size of the continental United States. The most important fact to realize about Brazil's rainforest is not that it is large, but that it is finite.

Brazil's Shrinking Rainforests

Although estimates of the extent and rate of deforestation in Brazil's Legal Amazon vary, the total felled area has been increasing exponentially. The Superintendency for Development of the Amazon (SUDAM) estimates that 115,000 km^2 were deforested during the 1966–1975 period (see table 1.1); other estimates go as high as 260,000 km^2 cleared by 1978 (W. E. Kerr, quoted by Myers 1980:128). The best prospects for resolving such discrepancies are by careful interpretation of available information from remote sensing.

Brazil has implemented remote sensing technology in its efforts to map and monitor the Amazon's forests and other natural resources (Brazil, RADAMBRASIL 1973–82; Hammond 1977a, b). One means used is side-looking airborne radar which, unlike air photos and satellite imagery, is not affected by the Amazon's notorious cloud cover. However, images from LANDSAT satellites offer a cheaper and more regularly renewed source of information

Table 1.1. SUDAM Estimates of Amazon Forest Clearing

	Amounts Cleared (km^2)	
Purpose	1966–1975	1976–1978
Cattle ranching	45,000	35,000[a]
Small farmers	35,000	
Highways[b]	30,000	
Timber harvesting	5,000	
TOTAL	115,000	

Source: C. Pandolfo, head of natural resources department of SUDAM (Myers 1980a:123–28).
[a] From figure of 80,000 km^2 for cattle ranching in the 1966–78 period. This 78 percent jump in three years would correspond to an exponential increase at 19.2 percent per year.
[b] Why so much area is needed for highway construction is unclear.

on forest clearing. Brazil's National Institute for Space Research (INPE) has its own satellite tracking facilities for receiving and processing LANDSAT data.[3] Later INPE reports, including LANDSAT image interpretation for the entire Legal Amazon, indicated that 7,771,175 hectares (ha) had been cleared by 1978, or 1.55 percent of this 497,552,700 ha area (Tardin et al. 1980:11). The difficulty of distinguishing second growth from primary forest on the images makes INPE clearing figures conservative. The best example is the Zona Bragantina, a 30,000 km^2 area near Belém that has been entirely cleared since the beginning of this century (Egler 1961; Penteado 1967; Sioli 1973:327). This area alone is larger than the 28,595 km^2 shown by LANDSAT data as cleared by 1975 in the entire Legal Amazon and is almost four times greater than the 8,654 km^2 shown as cleared by 1975 in the state of Pará (Tardin et al. 1980). "Very small" clearings are also not included in the LANDSAT estimates.

While the size of the cleared area in 1978 relative to the total size of the Legal Amazon is small, the areas cleared are increasing at an enormous rate: 169.88 percent between 1975 and 1978 (Tardin et al. 1980:11), corresponding to an annual increase of 17.66 percent. The form and rate of increase are far more important for the region's future than is the absolute area cleared at present.

LANDSAT data for clearing to 1980 are available for six of the nine states and territories of the Legal Amazon (see table 1.2). Clearing in three states (Rondônia, Mato Grosso, and Acre) appears to have followed a strong exponential trend over the 1975–1980 period, a tendency which becomes even more apparent when one considers the negligible clearing indicated by side-looking radar imagery for these areas in 1970 (Brazil, RADAMBRASIL 1973–1982).

A more detailed time series of LANDSAT deforestation data for a part of Rondônia suggests an exponential trend over the 1973–1978 period in this focus of intensive migration (Fearnside 1982a). Given the continued influx of migrants and investors into the region, it is reasonable to assume that the cleared area has increased substantially since the most recent images used for these estimates.

Concentration of Deforestation

The amount and rate of deforestation vary greatly among the regions of the Amazon. The greatest concentrations are located along the north-south swath of the Belém-Brasília Highway (BR-010) and its tributary roads, and

Table 1.2. LANDSAT Surveys of Forest Clearing in the Brazilian Amazon

State or Territory	Area of State or Territory (km²)	Area Cleared (km²)[a]			Percent of State or Territory Classified as Clear[a]		
		By 1975[b]	By 1978[b]	By 1980[c]	By 1975	By 1978	By 1980
Amapá	140,276	152.50	170.50	—	0.109	0.122	—
Pará	1,248,042	8,654.00	22,445.25	33,913.83	0.693	0.798	2.717
Roraima	230,104	55.00	143.75	—	0.024	0.062	—
Maranhão[d]	257,451	2,940.75	7,334.00	10,671.06	1.142	2.849	4.145
Goiás[d]	285,793	3,507.25	10,288.50	11,458.52	1.227	3.600	4.009
Acre	152,589	1,165.50	2,464.50	4,626.84	0.764	1.615	3.032
Rondônia	243,044	1,216.50	4,184.50	7,579.27	0.301	1.722	3.118
Mato Grosso	881,001	10,124.25	28,355.00	53,299.29	1.149	3.218	6.050
Amazonas	1,567,125	779.50	1,785.75	—	0.050	0.114	—
Legal Amazon (total)	5,005,425[e]	28,595.25	77,171.75	—	0.571	1.542	—

SOURCE: Fearnside 1984a.
[a] See text for explanation of why these values are underestimates.
[b] Tardin et al. 1980.
[c] Brazil, IBDF 1983.
[d] States not wholly within the Legal Amazon.
[e] Includes 27,138 km² of water surfaces (Brazil, IBGE 1982:28), following the practice of Brazil, IBDF (1983).

along the east-west path of the Cuiabá-Porto Velho Highway (BR-364) in Rondônia. Average clearing figures therefore give little idea of the human impact in areas of intensive settlement (Fearnside 1984a).

In Rondônia, a state in the southwest corner of the Brazilian Amazon, intense migration resulted in an increase in the human population at a rate of 14.6 percent per year between 1970 and 1980, a doubling time of less than five years (figure 1.2). The illusion of endlessness of the Amazon rainforest is a natural result of the human tendency to make linear rather than exponential mental projections of trends. Even in a country like Brazil where inflation has proceeded at rates ranging from 20 percent to over 200 percent per year, people are still continually surprised at price increases when they do their weekly shopping. One encounters older people who are at a loss even to explain the value of money from a former period—people who, for example, had bought a house for what would buy a bottle of Coca-Cola today. It is the same difficulty in internalizing exponential patterns that makes the limits of deforestation and population increase seem too remote to be taken seriously.

The economic and resource limitations potentially counteracting exponential trends at much increased rates of felling are unlikely to forestall for

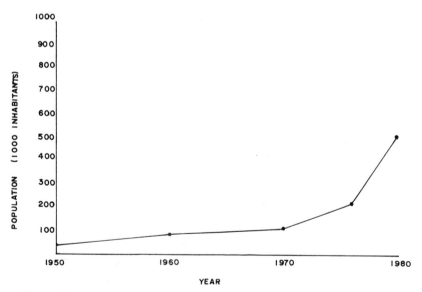

Figure 1.2. Population increase in Rondônia. Figures are from census data (Saunders 1974; Brazil, IBGE 1982:74; the 1976 intercensal estimate is by IBGE (Mesquita and Egler 1979:73)).

more than a few decades the clearing of all major forest areas in the region unless planning and policy enforcement are based on conscious decisions designed to contain the deforestation process (Fearnside n.d.(a)). Rainforest protection must be elevated to a major priority if such policies are to be effective.

Rainforest Protection

The rapid disappearance of rainforests in many tropical countries has encouraged some governments to focus attention on the limited nature of this resource. Steps to protect the last remnants of rainforest in countries such as Costa Rica (Myers 1979:142–48) are encouraging, although the trend continues toward rapid disappearance of rainforests worldwide. Deforestation of large areas is irreversible: the original forest cannot regenerate due to distance from seed sources, breaking of coevolved relationships, and other ecological changes (Gómez-Pompa et al. 1972).

Brazil, with more rainforest than any other country, maintains several types of reserves (Barrett 1980; Brazil, SEMA 1977; Brazil, IBDF 1979; Nogueira-Neto and Carvalho 1979; Padua and Quintão 1982), but they are small in relation to the size and diversity of forest types (table 1.3) and are rarely adequately protected against squatters (Brazil, INPA 1979). While the list of existing reserves reflects many significant gains in officially protected areas over the past decade, it also reveals the repeated pattern of both unofficial invasion and official abandonment of previous commitments to forest protection whenever land is desired for development purposes (table 1.3). Violations of reserves are likely to be even more frequent in the future as highway construction proceeds. In Rondônia, for example, government maps indicate proposed highways bisecting six Amerindian reserves and two biological reserves—one of these (the Guaporé Biological Reserve) to be criscrossed by three different roads (Fearnside and Ferreira 1984).

Worldwide Colonization of Rainforests

Governments see planned colonization of forest land as a solution to pressing problems of poverty, overpopulation, and inequities in land distribution. In addition to Brazil, other countries pursuing colonization in the Amazon basin include Bolivia, Peru, Ecuador, Colombia, and Venezuela (see Nelson 1973; IICA Programa Cooperativa para el Desarrollo del Trópico Americano 1972; Rundel 1983).

Table 1.3. Parks and Reserves in the Brazilian Amazon

Organ	Reserve Type	Name	State or Territory	Area (ha)	Year Created	Reference	Vegetation	Status
EMBRAPA	Research[a]	CPATU-1 (Altamira km23)	Pará	1	c.1975		High forest	
	Research[a]	CEPLAC/EMBRAPA	Amazonas	1	1973	Prance et al. 1976	High forest	
	Ecological area	Utinga	Pará	15.5		Pires and Prance 1977	High forest (5.7 ha) Igapó (4.8 ha) Várzea (5.0 ha)	
FAB	?	Aeronautica	Pará	180,000		Ayres 1977	High forest	Partly flooded by Tucuruí reservoir in 1985
FUNAI	Indian park	Araguaia-FUNAI	Goiás	1,395,000		Carvalho 1981	Cerrado and gallery forest	
		Aripuanã-FUNAI	Mato Grosso	1,258,322		Carvalho 1981	Transition forest, cerrado	Invaded (see Junqueira n.d. [1983]:58)
		Tocantins	Mato Grosso	624,994	1961	Pires 1978	Cerrado	
		Tumucumaque	Pará	2,978,500[b]	1961, 1968	Carvalho 1981	High forest	In path of Northern Perimeter Highway
		Xingú	Mato Grosso	2,642,003[c]	1961	Carvalho 1981	Cerrado	Partly invaded; partly rescinded for BR-80 Highway (see Bunker 1980a)

	Indian reserves (23)		various	5,619,174		Carvalho 1981	Various	Many are invaded
	Indian areas (110)		various	19,599,001		Carvalho 1981	Various	Many are invaded
	Indian posts (36)		various	4,828,377		Carvalho 1981	Various	Many are invaded
IBDF	National park	Amazônia	Pará	1,258,000[d]	1974	Carvalho 1981	High forest	Mostly disturbed; bisected by Transamazon Highway
		Araguaia-IBDF	Goiás	562,312	1959	Carvalho 1981; Padua and Quintão 1982	Cerrado	Bisected by highway (*Veja*, Dec. 22, 1982, p. 90; C. e C. 1983a:834)
		Cabo Orange	Amapá	619,000	1980	Carvalho 1981; Rylands and Mittermeier 1982	Mangrove and flooded savanna	
		Jaú	Amazonas	2,272,000	1980	Carvalho 1981; Rylands and Mittermeier 1982	High forest and *Campinarana*	
		Lençóis Maranhenses	Maranhão	150,000	1981	Carvalho 1981	*Cerrado*	
		Pacaás Novos	Rondônia	764,801	1979	Carvalho 1981; Brazil, IBGE 1980:44	High forest; *Cerrado*	Road built to park boundary, but construction halted 1982
		Pantanal Matogrossense	Mato Grosso	135,000[e]	1982	Pádua and Quintão 1982	Flooded savanna (*pantanal*)	

Table 1.3. Parks and Reserves in the Brazilian Amazon (*Continued*)

Organ	Reserve Type	Name	State or Territory	Area (ha)	Year Created	Reference	Vegetation	Status
	National park	Pico de Neblina	Amazonas	2,200,000	1961, 1979	Carvalho 1981; Pires 1978	Sub-montane forest	Road under construction bisecting park (C. e C. 1982:1237)
	Forest reserve	Gorotire	Pará	1,843,000	1961	Pires 1978	Transition forest	Disturbed, partially invaded, part ceded to cattle ranches
		Gurupí	Pará	203,000[f]	1961	Ayres 1977	Transition forest	
		Juruena Mundur-ucânia	Mato Grosso Pará	1,808,000 2,375[g]	1961 1961	Pires 1978 Pires 1978	*Cerrado* High forest	Invaded Completely invaded
		Parima	Roraima	1,756,000	1961	Pires 1978	Grassland and forest	
		Rio Negro Trans-amazônica[h]	Amazonas Pará	3,790,000 c. 90,000	1961	Pires 1978	High forest High forest	Partially invaded
	National forest	Caxiuaná	Pará	200,000	1961	Pires 1978	High forest	Being invaded for timber
		Tapajós	Pará	600,000			High forest	Partially invaded (*A Província do Pará* Dec. 4, 1974); being logged in forestry experiments to cover 140,000 ha

Biological reserve	Abufari	Amazonas	288,000	1982	C. e C. 1983b:248	High forest?	
	Jarú	Rondônia	268,150¹	1979	Carvalho 1981	High forest	Heavily invaded; highway planned which will bisect reserve
	Guaporé	Rondônia	600,000	1982	C. e C. 1983b:248	Flooded and high forest	Three roads projected crossing the reserve
	Lago Piratuba	Amapá	395,000	1980	Carvalho 1981; Rylands and Mittermeier 1982	High forest; coastal forest	
	Trombetas	Pará	385,000¹	1979	Carvalho 1981; Rylands and Mittermeier 1982	High forest	Mostly disturbed (D.C. Oren, pers. comm. 1982); mineral prospecting authorized (C. e C. 1982:1237); damaged by hydroelectric dam construction survey (C. e C. 1981:1504)
Research	Campina	Amazonas	90,000	1972	Brazil, INPA 1978	*Campina* and *campinarana*	
INPA	Dinâmica Biológica de Fragmentos Florestais (INPA/WWF-US)	Amazonas	Approx. 12,000	1980–1984	Lovejoy et al. 1983; R. O. Bierregaard, pers. comm. 1984	High forest	Reserves of 1–10,000 ha chosen such that they will be isolated as islands surrounded by cattle pasture as ranches continue clearing

Table 1.3. Parks and Reserves in the Brazilian Amazon (*Continued*)

Organ	Reserve Type	Name	State or Territory	Area (ha)	Year Created	Reference	Vegetation	Status
		Ducke	Amazonas	10,000	1958	Pires 1978	High forest	c. 15 ha cut for silvicultural experiments; some invasion
		Egler	Amazonas	630[k]	1968	INPA, pers. comm. 1981	High forest	
		Estação Experimental de Silvicultura Tropical	Amazonas	23,000	1972	Brazil, INPA 1978	High forest and *campinarana*	c. 100 ha cut for agricultural research
	Ecological research Reserve	Ouro Preto do Oeste	Rondônia	138	1983	Fearnside 1984g	High forest (c. 118 ha); secondary forest (c. 20 ha)	c. 10 ha of high forest understory damaged by fire in 1983.
SEMA	Ecological station	Anavilhanas	Amazonas	350,000		Brazil, SEMA 1977; Rylands and Mittermeier 1982	*Igapó* and permanently flooded forests (100,000 ha); high forest (250,000 ha)	
		Caracaraí	Roraima	80,560	1982	*A Crítica* June 1, 1982, p. 7	Savanna	
		Coco-Javaés	Goiás	37,000		Rylands and Mittermeier 1982	Seasonally flooded forest; savanna	
		Cuniã	Rondônia	100,000		IBRD 1981; Rylands 1984	High forest	Some squatters (G. L. Ferreira, pers. comm. 1983)

Iquê-Aripuanã	Mato Grosso	266,000		Brazil, SEMA 1977; Carvalho 1981	Transition forest; *cerrado*
Juami-Japurá	Amazonas	1,400,000	1983	*A Notícia* July 22, 1983, p. 3	High forest
Maracá-Roraima	Roraima	92,000		Carvalho 1981; Rylands and Mittermeier 1982	High forest; flooded forest; savanna
Maracá-Tipoca	Amapá	70,000		Brazil, SEMA 1977; Rylands and Mittermeier 1982	Mangroves; mudflats
Piriá-Gurupi	Pará	31,000		Brazil, SEMA 1977; Rylands and Mittermeier 1982	Mangroves; mudflats
Rio Acre-Sena Madureira	Acre	73,000		Carvalho 1981; Rylands and Mittermeier 1982	High forest
Serra das Araras	Mato Grosso	28,700	1982	*A Crítica*, June 1, 1982, p. 7.	*Cerrado*
Jutaí-Solimões	Amazonas	360,000	1983	*A Notícia* July 22, 1983, p. 3;	High forest; open palm forest

Ecological reserve

Table 1.3. Parks and Reserves in the Brazilian Amazon (Continued)

Organ	Reserve Type	Name	State or Territory	Area (ha)	Year Created	Reference	Vegetation	Status
SUDAM	Experimental forestry reserve	Curuá-Una	Pará	71,250	c.1957	Brazil, SUDAM	High forest	285 ha planted by 1974 of 550 ha planned for silvicultural plantations
	"Biological reserve" (forestry experiments)	Santarém Curua-Una km 62/64	Pará	1,200	1962–1964	J. Rankin and P. Fearnside fieldnotes 1978	High forest	65 ha cut for *Pinus caribaea* plantation; 3 ha cleared by neighboring ranch in 1978; logging and enrichment experiments throughout reserve

Note on reserves not included in the table: IBDF's Pedras Negras Biological Reserve in Rondônia (1,085,000 ha decreed in 1961) has been rescinded to make way for INCRA's Costa Marques colonization project. INPA's Aripuanã reserve in Mato Grosso (100,000 ha decreed in 1975; 140,000 ha as measured from RADAMBRASIL maps [Ayres 1978]) has been completely invaded by *grileiros* and squatters; the reserve has been abandoned by INPA and the research station there ceded to the Mato Grosso Secretariat of Agriculture. SEMA's Apiacás Ecological Station in northern Mato Grosso (500,000 ha) (Brazil, SEMA 1977) has been completely invaded and no longer exists (A. B. Rylands, pers. comm. 1984).

a Control areas for agricultural experiments.
b Also reported to be 2,560,000 ha (Goodland and Irwin 1975a:58), 3,084,500 ha (Brazil, RADAMBRASIL 1975:296). An area of 3,087,000 ha was decreed July 14, 1968 (No. 62.998), increasing reserve from 1,793,000 ha decreed July 25, 1961 (No. 51.043). See van Veltham (1980).
c Also reported as 2,200,000 ha (Pires 1978).
d None of this area is "regularized" (Pádua n.d. [1979]).
e Includes the area of the former Cará-Cará Biological Reserve (decreed 1961), variously reported as having 61,126 ha (Brazil, IBGE 1980-44; Carvalho (1981), 70,000 ha (Pádua n.d. [1979]), and 80,000 ha (Pires 1978).
f A 1,679,000 ha contiguous reserve in Maranhão (France 1975) apparently does not exist.
g Also reported to be 2,375,000 ha by France (1975:109).
h Altamira-Itaituba km 120.
i Originally decreed in 1961 with 1,085,000 ha (Pires 1978), the reserve was largely taken up by the PAD-Burareiro settlement scheme (IBRD 1981). The reserve was relocated and reduced in area in 1980–1981, apparently to allow cassiterite exploitation at the former site (Clay n.d. [1983]:18–19).
j Originally 1,258,000 ha (Brazil, IBGE 1979:44).
k Previous 750 ha size (Pires 1978) reduced: land lost in 1980.

The phenomenon is pantropical. Most recently, Indonesia has massively accelerated a longstanding program to relocate populations from Java and three other densely populated islands to rainforest areas on less densely populated islands (Jones 1979). Indonesia's Third Five-Year Development Plan (*Replita* III), covering the 1979–84 period, calls for moving 500,000 families, or about 3 million people, during the plan (Ross 1980). The plan calls for continued movement until a total of two million families, or 12–15 million people, is reached (Myers 1980a:73–74).

While carrying capacity is generally not considered in planning colonization projects, it has begun to enter into the practical side of planning in a few tropical countries. The government of Fiji is reportedly making good use of a UNESCO study which included carrying capacity estimates of some areas (Bayliss-Smith 1980). In Papua New Guinea, where an unusually large number of academic studies have been done involving carrying capacity, the National Planning Office has included rough calculations of carrying capacity for shifting cultivation in considering a proposed transfer of population from densely populated highland portions of the province of Chimbu to lowland rainforest areas (Simpson 1975). The formula for estimating carrying capacity under shifting cultivation used in Papua New Guinea and in many other studies was originally devised with the practical problems of African agricultural development in mind (Allan 1949). Some have suggested that shifting cultivation formulas for carrying capacity estimation would be appropriate for use in colonization areas in the Brazilian Amazon (Morais 1974a; Valverde 1979a:248). Nevertheless, the bulk of planned settlement taking place in tropical countries, including Brazil, is done without any explicit consideration of carrying capacity.

The Brazilian government's decision to promote rapid settlement in the Amazon through road construction and colonization programs has contributed substantially to the pressure on rainforest areas. An examination of the motives for this decision, the colonization programs, and some of the other occupation and exploitation patterns being promoted and considered will make clear the force and extent of this impact.

Brazil's Colonization of the Amazon

Motives

The drive to settle the Amazon is a continuation of earlier Brazilian government efforts to develop the interior, such as the construction of a new national capital at Brasília (1956–1960) and the building of the Belém-Brasília Highway (1957–1960). Following the entry of the present government in 1964 the efforts were redoubled, with the improvement of the Belém-Brasília Highway for year-round traffic in 1964, the creation of the Superintendency for Development of the Amazon (SUDAM) in 1966 and the Manaus Free Trade Zone (SUFRAMA) in 1967 and the opening of the Cuiabá-Porto Velho Highway connecting Mato Grosso with Rondônia in 1968. Colonization projects were also undertaken in the Amazon during this period (Tavares et al. 1972). Not until 1970, however, did the great push begin with the announcement of the National Integration Program (PIN) and the building of the Transamazon Highway.

During an emotion-charged speech to drought victims in Recife in northeastern Brazil, on June 6, 1970, President Medici promised to do something about their plight.[4] Ten days later, plans for the National Integration Program and the Transamazon Highway were announced (Decree Law 1106). Bids were solicited from contractors for constructing the highway only two days later (Rebelo 1973:87). The highway was to stretch from Recife and João Pessoa on the Atlantic coast at the "bulge" of Brazil to the Peruvian border near Cruzeiro do Sul in Acre: a total length of 5400 km, about 3000 km of which is in Amazonia (figure 1.1).

Official pronouncements emphasized social factors as the justifications for programs to develop the Amazon. Poverty in northeastern Brazil was the immediate issue, especially following the drought of 1970. Overpopulation was stressed as a root cause, although other problems such as extreme inequality in land tenure distribution were not addressed. Transfer of poor northeasterners to colonization areas along the Transamazon Highway was put forward as the solution. Sixty-five percent (de Arruda 1972:5.9) to 75 percent (Brazil, INCRA 1972a:1) of the colonists were to come from the Northeast, although in fact only 30 percent of those settled in the Altamira area by the end of 1974 actually came from that region (Brazil, INCRA 1974).

A second official goal was economic benefit for the country. Through agricultural credit and technical assistance the colonists were to produce a surplus for export to other parts of the country or to other nations. The road would also make timber and mineral resources accessible. One government pamphlet speaks of the "ecstatic panorama" of mineral wealth along the highway (Brazil, INCRA c.1972:6). J. M. G. Kleinpenning (1979:38) suggests that the initial stress laid on social motives was merely a "useful political maneuver" (see also Ianni 1979), and that other motives, especially the achievement of economic growth, were more important.

A third motive is strategic geopolitical considerations (Kleinpenning 1979; Lima 1973). The fear that Amazonia is never far from invasion by foreigners is a recurrent one in Brazil. The publication of four editions since 1957 of Artur César Ferreira Reis' 1972 A *Amazônia e a Cobiça Internacional* (Amazonia and International Covetousness) testify to this. A casual suggestion attributed to a speech made by Harrison Brown, author of *The Challenge of Man's Future* (1954), that the population problems of India might be "solved" by moving populations to Amazonia (Reis 1972) made headlines in Brazilian newspapers. In his book *Principles of Political Economy*, economist Kenneth Boulding suggested moving 200 million Asians to Amazonia, contributing to concern in Brazil (Pinho Filho 1979:119). Rapid occupation of the area by Portuguese-speaking Brazilians was seen as the best defense against this potential influx of foreigners. One of the objectives proclaimed for the National Integration Program was to "mark, by the presence of Brazilian men in Amazonian lands, the conquest for themselves and for their country, of that which always belonged to them, so that no one would ever dare to contest them on this objective" (Brazil, INCRA 1972a:1). More than the possibility of invasion from outside, Kleinpenning (1979:38) stresses the strengthening of Amazonia as a base for Brazil's pursuing what he calls its own "sub-imperialistic motives" (1979:38; see also Tambs 1974).

Other motivations include the regime's desire for prestige, both international and national, and the pressures of national politics (Kleinpenning 1979:4). "Spectacular and exciting actions of government, such as the active development of the Amazon region and enthusiasm for it among larger sections of the population, can result in political conflicts being felt less severely for a time and in attention being temporarily diverted from such matters as lack of political freedom, torture, and social injustices" (Kleinpenning 1979:4). In 1970, when these problems were at a high point, promotion of the Transamazon Highway in São Paulo and other places through billboards, advertisements in buses, T-shirts, and so forth, went far beyond the announcements needed to recruit prospective colonists. The flood of migrants to São Paulo from the northeast was seen by the public as the source of all urban ills, making a simplistic but highly visible alternative popular (Katzman 1976:456). Political stability is an important goal of promoting enthusiasm and euphoria for projects such as the Transamazon Highway (Kleinpenning 1979:4).

The same solution had been considered by previous governments. In *The Hungry Planet*, published five years before the drought of 1970 and the Transamazon Highway, Georg Borgstrom (1965) wrote: "Another catastrophy [drought] scorged this region [northeast Brazil] in 1957–58 when 70 percent of the crops were lost. . . . Each time this happens, agitators get new grist for their revolutionary mills. . . . One Brazilian secretary of finance came to talk the United States Government into financing the transfer of this population to the Amazonas" (p. 317). The idea of transferring northeasterners to the Amazon after the recurrent dips in the short-term capacity of the Northeast to support a human population is not new: Emperor Dom Pedro II offered drought victims free passage to the ports of Belém and Manaus following the drought of 1877 (Morais et al. 1970:115).

It is fruitless to search for a single motive for initiating the highway construction and colonization programs in the Amazon. Many reasons undoubtedly contributed to the attractiveness of this course of action. The questions of how colonists should be selected and colonization programs pursued hinge on the sometimes conflicting implications of the various motives. Many of the motives that contributed to decisions regarding past programs will be important as future programs are planned and executed.

Colonization Programs

Colonization Prior to 1970

Nonindigenous settlers have been colonizing the Brazilian Amazon since the sixteenth century. The major settlement pattern in most of the region has been sparse dotting of the riverbanks with villages or individual holdings of *caboclos*,[5] who have traditionally supported themselves with subsistence plots of annual crops, mainly manioc, and by fishing, hunting, and extracting forest products such as Brazil nuts *(Bertholetia exelsa)* and rubber *(Hevea brasiliensis)* (Wagley 1976; Moran 1974).

Intensive settlement took place along the railway constructed from Belém to Bragança at the end of the nineteenth century. Agricultural colonies were established to supply the rubber boom city of Belém (de Camargo 1948; Sioli 1973; Penteado 1967). The dense population in the "Zona Bragantina" practiced an accelerated shifting cultivation with inadequate fallows, resulting in soil exhaustion and declining crop yields (de Camargo 1948; Egler 1961; Ackermann 1966; Sioli 1973, 1980). The decline of agricultural productivity, with subsequent population emigration and abandonment of much of this 30,000 km^2 area to second growth, is an example of what can occur if colonization programs in the Amazon exceed carrying capacity (see Penteado 1967). The problem of "fixing the man to the land," considered acute in the Zona Bragantina (Penteado 1967), can arise from exceeding carrying capacity, since if agriculturalists cannot support themselves over the long term they will become "unfixed" and leave the area.

Other types of colonization have involved the establishment of more specialized groups, such as the colonies of Japanese immigrants who have concentrated on growing black pepper *(Piper nigrum)* and vegetables (see Fearnside 1980a). Rubber plantations have also formed the basis for past colonization, aside from the thousands of scattered gatherers of rubber from wild trees in the forest. The Ford Motor Company established plantations at Fordlândia on the Tapajós River between Santarém and Itaituba in 1926 (later abandoned), and 100 km downstream at Belterra in 1934 (later turned over to the Brazilian government when failing as an economic venture) (Sioli 1973).

The colonization of the Brazilian Amazon through planned small farmer settlements, as well as other types of holdings along new highways, dramatically accelerated in 1970 with the announcement of the National Integration Program (PIN).

The National Integration Program

The advent of the PIN led to the creation of a new government organ for colonization—INCRA, the National Institute for Colonization and Agrarian Reform. INCRA set out to colonize areas in Amazonia by the establishment of PICs (Integrated Colonization Projects) and PADs (Directed Settlement Projects). The colonization of the Transamazon Highway is divided into three separately-administered PIC areas: Marabá, Altamira, and Itaituba, all established in the early days of the National Integration Program. There are also five PICs in Rondônia (Brazil, INCRA 1972b; Valverde 1979b).

Directed Settlement Projects (PADs), a more recent form of planned colonization, do not provide as much supportive infrastructure for the settlers as do the PICs. The two PADs in Rondônia were established in 1974 and 1975. Another PAD is currently underway in Acre, a state in western Amazonia bordering on Peru and Bolivia. The high cost of implanting and administering the PICs, together with the fact that colonists have shown themselves to be more than willing to migrate to the Amazon without the inducement of services offered by these projects, undoubtedly explains the change to PADs. Colonists to be accommodated in PADs are also expected to have greater financial resources than those in PICs (Neves and Lopes 1979:87).

The colonization programs in the three PICs of the Transamazon Highway were to settle 100,000 colonist families over a period of five years.[6] The colonization areas were laid out in accord with what INCRA called the "philosophy of rural urbanism" (da Cunha Camargo 1973).[7] Colonists' lots are served by a hierarchy of three types of planned centers: the agrovila, a small village of about 50 houses laid out in a rectangle, the agrópolis, a town expected to service 8 to 10 (da Cunha Camargo 1973:16) or 22 (Brazil, INCRA 1972a:67) agrovilas, and the rurópolis, a city to have a population of around 20,000 (Brazil, INCRA 1972a:67). In reality only 29 agrovilas, two agrópoli, and one rurópolis were ever actually built (Kleinpenning 1979:22). By December 1974, only 5,717 families had settled on the highway (Smith 1976a, b) far short of the goal of 100,000 by 1976. New settlement of small colonist lots on the Transamazon Highway has remained at a virtual standstill since that time, although regularization of squatter claims in an area east of the Altamira PIC began in 1982.

Each colonist brought to the colonization areas by INCRA at the outset of the National Integration Program received a 100-hectare lot called a *lote* for a nominal price which was to be paid over a twenty-year period. A three-year grace period was given before beginning payment and 7 percent annual interest was charged on the unpaid balance. The usual practice in Brazil of charging monetary correction to adjust for the country's astronomical inflation was not applied to the land-purchase loans.

In most of the Altamira PIC, and in a part of the Marabá PIC, lateral roads called *travessões* run perpendicular to the Transamazon Highway at intervals of five kilometers. Most lateral roads extend the range of small colonist settlement to about twenty km on each side of the highway. Lots fronting on the main highway are 500 meters wide by 2,000 meters deep, while those in lateral roads are 400 by 2,500 meters. Lots are grouped into units of 10 to 70 lots called *glebas*, each *gleba* occupying approximately five kilometers of one side of the main highway. In the Altamira area, agrovilas are spaced every ten kilometers along the main highway, and at depths of ten kilometers in the lateral roads. Most colonists with roadside lots had houses built by INCRA on their farms, while those with lots on the lateral roads had houses in one of the agrovilas (figure 1.3).

The Altamira area is by far the largest of the three Transamazon Highway PICs. The portion of the Altamira area lying in the Altamira-Itaituba section of the highway has approximately 3120 colonist families, or about 59 percent of the total for the Transamazon Highway. This area covers a strip ex-

Figure 1.3. Agrovila Grande Esperança, the planned agricultural village where I lived during two years of fieldwork (1974–1976). My house is on the right. (Photo by J. M. Rankin, 1975).

tending from 12 to 245 km west of Altamira, with one 30 km break for a forest (timber) reserve.

Many colonists leave their houses in the agrovilas in favor of more rustic accommodations (figure 1.4), closer to their fields and barnyard animals. Lack of services such as schools, water, and health clinics in many agrovilas reinforces this pattern. Agrovilas on the main highway have much better urban services than those in the interior.

The emphasis of colonization programs has changed regularly since the initial launching of the National Integration Program. A major policy change occurred in 1974, when colonization by individuals on 100 ha lots was deemphasized in favor of "colonization" by large corporations (*O Estado de São Paulo* May 24, 1974, p. 11). These large corporations, mainly cattle ranching operations, included both giant internationals and hundreds of Brazilian investors from urban areas in southern Brazil. Smaller ranches were sold by INCRA in the area beyond the strip of small colonist settlement on the Transamazon Highway. Ranches of 3,000 ha each were sold in the area about 150 km west of Altamira beginning in 1974. Similar sales were made in Marabá and Rondônia. Beginning about 1977, 500 ha ranches, called *glebas*, were sold in strips about 30 km wide beyond the edges of the small colonist settlement between 12 and 85 km west of Altamira. Land is sold

Figure 1.4. Many colonist families have abandoned houses in *agrovilas* in favor of more rustic accommodations on their lots (Gleba 16, Lot 60, 1975).

through the process of *licitação*, soliciting sealed tenders for individual parcels with a minimum bid of 2 percent of the official minimum monthly wage per hectare, or about US $1.08/ha.

Colonization schemes by private cooperatives have been encouraged, to avoid the many inefficiencies inherent in government colonization. Such a private colonization scheme was to be undertaken in an area south of the Transamazon Highway about 110 km west of Altamira, on the left bank of the Iriri River. It was announced in 1976 that this area would be colonized by COTRIJUI (Cooperativa Tritícola Serrana), a cooperative of small farmers from the extreme south of Brazil, with each of 2,000 families receiving a 200 ha lot (*O Estado de São Paulo*, May 21, 1976). This plan replaced an earlier one to colonize the area through INCRA (Hirano 1974). The cooperative has not been able to begin the proposed colonization project due to hostile Arara Indians living in the area. FUNAI (the National Indian Foundation) continues its efforts to dislodge the tribe, which on February 22, 1981, made its first peaceful contact with Luso-Brazilian culture since the Transamazon Highway was built through the tribe's lands eleven years previously (*Veja*, March 11, 1981, pp. 72–76).

Spontaneous Settlement

The largest part of the settlement in the Amazon today, as in the past, is done with no government or other planning whatsoever. Unplanned colonization by squatters, the traditional means of settlement, has engendered many bloody fights throughout Amazonia between squatters and either landowners holding documents for legal ownership, or the more feared *grileiros*, speculators who contract thugs (*jagunços* or *pistoleiros*) to drive small farmers off the land they occupy. Grileiros obtain official (often fraudulent) documents allowing sale of this land to ranching interests (Bunker 1980a; Martine 1979, 1980; Martins 1980; Schmink 1982; Wood and Schmink 1979). Settlement of the Belém-Brasília Highway took place in this way, with land that had been cleared by small farmers without documents later taken over by large ranchers (Valverde and Dias 1967:276). Sometimes the process worked in reverse, with absentee investors losing to squatters land they had bought (Sanders 1971).

Official colonization programs cannot cope with even a small fraction of the influx of new migrants to the Amazon: of the 8,000 persons who were registered crossing a government checkpoint entering Rondônia from Mato Grosso in September 1979, 7,000 settled in Rondônia. While the flow averages over 2,000 families per month over a twelve-month period, it is reduced but not stopped during the rainy season, with 3,671 persons in 900 families entering during January and February 1980 (*A Crítica* [Manaus], March 12, 1980, section 1, p. 7). The flow of migrants increases yearly, the January–February flux at the checkpoint nearly doubling to 6,435 persons in 1981 (Modesto 1981:26). Migration figures from the government checkpoint are substantially lower than actual flow, since many migrants pass uncounted. Through 1977 INCRA had settled 12,660 families on 2,732,550 ha of land in Rondônia, as compared with a total of between 5,000 and 6,000 families on 100-hectare lots in the three colonization areas of the Transamazon Highway. Rondônia has the most INCRA-planned colonization projects of any part of the Amazon, and lots in these projects are fully occupied, with the possible exception of one PIC (Sidney Girão), reportedly only half full due to its location in a remote and less fertile area where it was installed "apparently with the intent of occupying areas near the Bolivian border" (Mueller 1980). The majority of new migrants settle in areas outside INCRA projects in the traditional pattern of spontaneous squatter settlements.

Agriculture on the Transamazon Highway

The kind of market-oriented pioneer farming done in the Transamazon Highway Colonization Area is largely based on annual crops, with upland rice being the most prominent cash crop (figure 1.5). Maize, beans, and manioc are planted as cash crops on a smaller scale. The approach is usually land-intensive, with labor and especially capital inputs minimized. Financing has enabled many colonists to expand the size of their plantings by hiring supplementary labor. Colonists as a rule do not practice shifting cultivation with fallow periods that renew soil nutrients lost from cropping. Many plan to convert their land into pasture or perennial cash crops such as black pepper or cacao. Since land is being cleared for annual crops faster than

Fig. 1.5. Typical rice fields in newly cleared land. Rice alone on left, rice interplanted with maize on right. (Gleba 18, Lot 24, 1974).

available capital and labor permit it to be converted into perennial crops, much of the land used for one or two years under annual crops is either left to develop *capoeira* (second growth) or is sown as pasture. A significant amount of pasture has been planted by colonists, often remaining for several years with no fencing or livestock. Considerable variation exists among colonists in the type of farming employed, some of which appears to be explained by the colonists' backgrounds (Moran 1975, 1976, 1979b; Fearnside 1980b; Fleming-Moran and Moran 1978). A second wave of colonists has been arriving on the Transamazon Highway and either buying lots from the original colonists settled by INCRA or buying abandoned lots. Original colonists were supplanted first in the roadside lots, such as those with large areas of pasture in figure 1.6, but since 1976 the process has progressed into the lateral roads. There are now virtually no abandoned lots remaining for which the "right of possession" has reverted to INCRA, although some lots are always held by speculators who do not work the land. Newcomer colonists arrive with greater capital resources than the original colonists and exhibit different behavior patterns in land use allocation.

Farming by colonists on the Transamazon Highway falls into an annual cycle in accord with the wet and dry seasons. The agricultural year is best

Figure 1.6. Typical view of the Transamazon Highway showing roadside lots bought by newcomer colonists for planting pasture (Gleba 20, Lot 1, 1975).

conceptualized as running from July 1 to June 31. The *broca*, or underclearing, of virgin forest begins in July during the dry season and continues into August. Underclearing removes forest understory and vine components in preparation for felling. Felling of the larger trees begins after underclearing is completed and usually lasts until about October, although occasionally it can last until as late as January. Second growth can be cut later than virgin forest and still be burned before rain makes burning impossible. Unlike an Asian monsoon, the rainy season in the area begins gradually with slightly more rain falling every week until the season is finally underway.

After felling or second-growth clearing the vegetation is allowed to dry for two to eight weeks. Burning can be accomplished in a single day if weather and timing have been favorable, but can extend over weeks and even months if a colonist is trying to burn during short rainless breaks in a wet year (as was the case in 1973), or if burning has been left until too late in a "normal" year.

Following burning a large amount of unburned material is often left in the field. If the burn has been very poor, the colonist simply abandons the field for that year; otherwise a *coivara* is done or piling up unburned vegetation in heaps for a second burning. The amount of effort expended on this extremely laborious task varies with the amount of time left until the rains begin, the quality of the burn, and the crop to be planted next. Rice fields are usually the best cleaned, followed by manioc and then maize fields.[8] Fields planted directly to pasture have no *coivara* done at all, including fields burned too poorly for any other crop. The subsequent planting, weeding, and harvesting practices depend on the individual crops planted.

Land Use in the Amazon

Small farmers, both within and outside of planned settlement schemes, are being replaced by other types of family and corporate enterprises. Other areas are settled directly by large enterprises engaged in cattle ranching, plantation agriculture, and forest exploitation.

Ranching

Cattle ranching has been the most widespread form of land use in the Brazilian Amazon, rapidly altering the landscape in the region's accessible portions. Large cattle ranching enterprises have been established along the Belém-Brasília Highway and associated branch roads in eastern and southern Pará, and the Amazonian parts of Maranhão, Goiás, and Mato Grosso states. These include Liquigas (678,000 ha), Volkswagen (139,000 ha), Armour-Swift/Brascan/King Ranch (72,000 ha), and many other large corporate investors (see Goodland 1980a, b; Myers 1980a, b). Somewhat smaller ranching enterprises, usually with holdings of less than 10,000 ha, predominate in other regions of rapid deforestation such as Acre, Rondônia, and central Pará, although a few holdings in these areas are far larger.

Much of the upland rainforest area converted to pasture has become degraded, or invaded by second growth. Estimates of the extent of pasture degradation vary widely (table 1.4). Most upland pastures become choked with woody vegetation after five to seven years (dos Santos et al. 1980; Fearnside 1979a; Toledo and Serrão 1982). Guinea grass (*colonião* : *Panicum maximum*), which occupies 85 percent of the area of planted upland pasture (Serrão and Falesi 1977:54), is particularly susceptible to invasion due to its bunchy habit and poor reseeding. Soil compaction and declining phosphorus levels slow pasture growth. On the Transamazon Highway, the quality of pastures declined visibly throughout the 1973–85 period. Overgrazing, previously rare in the colonization area, has been common since about 1981.

Table 1.4. Estimates of Extent of Pasture Degradation in Amazonia

Percent of Pasture Area Degraded	Date of Estimate	Location	Method	Reference
17	1979	Amazon Region	Unstated[a]	Serrão 1979
20	1978	Brazilian Amazon	Unstated[a]	Serrão et al. 1979:202
17–24	1980	Amazon Basin	"Unofficial reports . . . from officials and cattle farmers"[a]	Toledo and Serrão 1979:292
44	1975–1976	Barra de Garças and Luciara, Mato Grosso	LANDSAT	Tardin et al. 1978:24
50+	1980	Paragominas, Pará	Unstated[a]	Hecht 1981:83
54	1977–1978	Paragominas, Pará	LANDSAT	dos Santos et al. 1979:100

[a] Probably based on informal impressions from ground-level observations during travel in the region.

In addition to planted pastures, most of the 15 million hectares of "natural" upland grasslands and 1.5 million hectares of *várzea* (floodplain) grasslands in the region are used for cattle. Most "natural" grassland areas are found in the federal territories of Roraima and Amapá, and in the area of Humaitá in the southern part of the state of Amazonas, while the periodically flooded *várzea* grasslands are along the Amazonas (Lower Amazon) and Solimões (Upper Amazon) rivers, and in coastal areas of the "fresh-water ocean" near Belém.

The rush to establish cattle ranches in the Brazilian Amazon has been speeded by generous government inducements in the form of tax incentives and loans with negative interest rates in real terms (after considering inflation) (see Bunker 1980a; Fearnside 1979b, 1983a). Loans given by the Banco da Amazônia, S.A. (BASA) after project approval by the Superintendency for Development of the Amazon (SUDAM) undergo adjustments for inflation at official rates, invariably lower than actual inflation. Loans have two-year grace periods before payments begin; the original repayment schedule of seven years has been reduced to five years for more recent loans. Part of the tax owed on income earned elsewhere in Brazil by companies undertaking the projects can be applied toward capitalizing ranching enterprises. An additional program provides direct subsidies to approved projects. Tax incentives and other subsidies accounted for 72 percent of the funds invested in Amazonian cattle ranches in 1977 (Kohlhepp 1980:71). In 1979 the government declared a moratorium on SUDAM approval of fiscal incentives for new ranching projects in those parts of Amazonia classified as rainforest, although incentives for projects already underway continue. The parts of the Legal Amazon with *cerrado* (scrubland) vegetation, as well as a large area classified as "transition" forest, continue to receive incentives for new projects.

Between 1967 and 1978 SUDAM approved 335 ranching projects in the Legal Amazon, covering an area of 7,887,169 ha (de Almeida 1978:28). Fiscal incentives, while important, are not the only force leading to deforestation for cattle ranches. A survey of 445,843 ha of clearing in rainforest (Tardin et al. 1978:19) indicated that nearly half was cleared without fiscal incentives.

Land Speculation

The appreciation of land values provides an additional powerful incentive for the rush to pasture, motivating investors to undertake ranching operations in Amazonia despite poor agronomic prospects. The ranching operations themselves produce a meagre amount of beef, and there is little reason to expect that production can be economically sustainable over an extended period. A SUDAM survey of twelve large ranches in Pará found annual net profits to be only about US $3/ha after ten years, and less than US $6/ha after twenty years (Serete S.A. Engenharia and Brazil, Ministério do Interior, SUDAM 1972:13–23). A number of ranches surveyed were operating at a loss.

The size of ranches has a marked effect on land values, which are influenced by the varied enforcement of conflicting laws and government policies in different parts of Amazonia. In Rondônia, INCRA has enforced government policy of not distributing public lands *(terras devolutas)* in parcels larger than 2,000 ha.[9]

Developers of large cattle ranches consequently have preferred Mato Grosso and Acre where this law has not been enforced. This policy has resulted in land values in Acre six times higher than those in Rondônia according to one report (Théry 1976:96). SUDAM has confirmed its preference for large ranching projects by requiring a minimum of 25,000 ha to qualify for incentives (Cardoso and Müller 1978:162).

Increase in land value can make even a marginal ranching operation highly profitable in the long run, provided title to the property can be obtained and held. Qualifying for a piece of paper with a few signatures and rubber stamps can thus add much more to a property's value than the production of beef cattle. Clearing land and planting pasture is one way of maintaining claim to the land and qualifying for the *título definitivo* (definitive title).

Speculators have often opted for cattle ranching because of its low cost per hectare for implantation as compared with other uses. Increases in the prices of pasture land are spectacular by any standard. In Amazonian Mato Grosso, real prices of pasture lands (after discounting inflation) increased at an *annual* rate of 38 percent during the 1970–75 period (Mahar 1979:124). These gains could be realized without any agricultural production whatsoever. Furthermore, the resale value of the land has a tendency to become detached from the land's theoretical value in terms of expected future production (Found 1971:24), to the extent that it is treated as a commodity like gold bullion or rare stamps, whose value does not derive from actual usefulness as a production input. Indeed, continued productivity for these pastures is highly unlikely (Hecht 1981; Fearnside 1979a, 1980c). The root of the motivation to bid the price of land far above its value for production is undoubtedly desire for shelter from Brazil's approximately 200 percent per year inflation. Recent land speculation in Amazonian pasturelands probably could be counted among the most profitable investments on earth, giving speculators a powerful motive for rapid occupation and implanting of pasture.

Perennial Crop Plantations

Monospecific stands of a number of crops are becoming increasingly important in development of Amazon *terra firme*. Present areas are small relative to both the Amazon's total area and to land being planted in pasture, but plans for expansion continue. The limited capacity of world markets to absorb great increases in production guarantees that areas in perennial crops will remain small relative to the total.

Rubber is a major priority. Brazil, once the world's major rubber source, was forced to import 47 percent of its natural rubber needs in 1982 (A *Critica* [Manaus] September 24, 1982, p. 7). Rubber is promoted by a special government organ, SUDHEVEA, under a program called PROBOR, as well as through the fiscal incentives of SUDAM. By 1979, 15,000 of the 19,000 ha of rubber planted under the PROBOR program were considered satisfactory, with the target for the program set at 40,000 ha (Morais 1979).

Cacao *(Theobroma cacao)* plantations are financed both for small farmers in some of the planned colonization projects of the Transamazon Highway and Rondônia, and for larger landowners in neighboring areas where 500–2,000 ha parcels have been sold through *licitação* (closed tenders). A government plan undertaken by CEPLAC (the government cacao promotion and research organ) calls for financing 200,000 ha of cacao in the Amazon over a fifteen-year period (Alvim 1977a:350). In Rondônia, the area in cacao is estimated to have increased from 17,528 ha in 1978 to 33,528 ha in 1980 (Brazil, CEPA-RO 1980: quadra 20). The world market price for cacao has been falling in real terms since 1977, a trend that the World Bank expects to continue through 1990 (Skillings and Tcheyan 1979; IBRD 1981). Future expansion can therefore be expected to slow sharply.

Black pepper *(Piper nigrum)* has been planted without government inducements in several areas of Pará and Amazonas. More recently, government financing has been made available for black pepper, mostly to colonists in the Transamazon Highway area. Land under pepper in Pará increased from 5,674 ha in 1973 to 8,197 ha in 1976 (Homma and Miranda Filho 1979:18). A fungal disease *(Fusarium solani* f. *piperi)* has forced the abandonment of older plantations and caused planters to migrate to new locations (Fearnside 1980c).

African oil palm (*Elaeis* spp.) may become more widespread as a plantation crop in coming years. In 1981 a French firm began a commercial-scale plantation near Tefé, in the state of Amazonas, but on a smaller scale than the 33,000 ha originally contemplated (de Almeida 1977). Meanwhile, an experimental plantation of France's *Institute de Recherche pour les Huilles et Oleaginneaux* (IRHO) had 1500 ha planted near Belém by 1977, with plans to expand with an additional 3,500 ha (de Almeida 1978:31). This experimental work has led to a commercial enterprise, Dendê do Pará S/A (DENPASA), with 2,500 ha producing by 1979 (Muller 1979).

Guarana (*Paullinia cupana*), a sapindaceous woody climber used in a soft drink, is a native plant grown on relatively small plantations. It is especially common in areas of the state of Amazonas near Maués and between Manaus and Manacapurú. SUDAM incentives have been approved for processing plants (de Almeida 1978:31). Government plans to finance extension of guarana culture in other areas, including the Transamazon Highway, have so far remained on paper.

Coffee plantations have been financed in Rondônia for colonists with 100 ha lots in planned colonization projects. The area under coffee in Rondônia is increasing rapidly, estimated to have jumped from 20,091 ha in 1978 to 32,638 ha in 1980, with the area already in production rising from 6,630 ha to 19,567 ha in the same period (Brazil, CEPA-RO 1980: Quadra 18). A much smaller number of colonists have planted coffee in the Altamira area of the Transamazon Highway in Pará. As in the case of cacao and other perennial crops, the limited capacity of world markets to absorb vastly increased quantities of the product can be expected to slow the expansion of planted areas long before any significant part of this region's vast area is converted to these crops.

Silvicultural Plantations

Silvicultural (tree) plantations have been increasing in the Amazon and are being encouraged as an appropriate form of development for the region (Alvim 1977a, b; 1978a, b, c). Plantations could supply the country's wood and paper needs from a much smaller area than would be the case if natural forest were used. They could have an especially important effect in reducing pressure for rainforest felling if installed in previously cleared areas.

Hardwood species have so far been planted only in experimental settings, but fast-growing species for pulp, plywood, and sawlogs have been planted in an increasing number of commercial plantations. ICOMI, a manganese mining operation in which Bethlehem Steel has a 49 percent interest, has planted 20,000 ha of Carribean pine *(Pinus caribaea)* in Amapá Territory. Georgia Pacific has been planting *Pinus caribaea* on their 500,000 ha property near Portel in the state of Pará (Cardoso and Müller 1978:161).

Best known are the plantations of Companhia Florestal Monte Dourado, S.A. (formerly Jari Florestal e Agropecuŕia Ltda.), better known simply as "Jari." Jari, an estate claiming 1.6 million hectares, was originally developed by the shipping magnate Daniel K. Ludwig. In 1982 a controlling interest in the estate's silviculture and mining projects was sold to a group of Brazilian firms now numbering twenty-three. The property straddles the Jari River, the tributary to the lower Amazon forming the boundary between Pará and Amapá. The first plantations were installed in 1969, and by 1983 a total of 90,600 ha were planted in monospecific silvicultural stands, including 10,000 ha that had ceased to be managed. *Gmelina arborea*, the species initially intended to be the sole species, grew poorly on sandy soils occupying much of the estate, leading the management to convert some areas to *Pinus caribaea* and later to species of *Eucalyptus*.

Gmelina arborea is a tree of Asian origin known for rapid growth (Palmer 1973). *Gmelina* at Jari has been attacked by a variety of insects and diseases (Fearnside and Rankin 1980). In 1974 an unidentified lepidopteran larva rapidly defoliated 300 ha of *Gmelina*. Subsequent outbreaks have so far been restricted to smaller areas, but this and other pests continue to infest the plantations every year. A fungus *(Ceratocystis fimbriata)* has appeared in several parts of the *Gmelina* plantations since 1976 (Muchovej et al. 1978). Trunk cancres caused by this fungus eventually kill the trees. Chemical control of the fungus is presently considered impractical and uneconomic. The disease is now inflicting severe losses, both by direct damage to *Gmelina* trees and indirectly by obliging Jari to adopt costly management changes such as harvesting the trees at an earlier age and substituting *Gmelina* with the less valuable *Eucalyptus deglupta* on some soils that could have produced *Gmelina*.

Pinus caribaea var. *hondurensis* has been planted on a commercial scale at Jari since 1973. The leaf cutter ant (*Atta* spp.), the principal pest of *Pinus*, is controlled with pesticides (Ribeiro and Woessner 1978). *Atta* can kill *Pinus* seedlings up to two years old and reduce growth. Losses of *Pinus* at Jari are not economically unacceptable when balanced against the rate of growth during the first harvest cycle. Eventual need for fertilizers and other inputs, as well as the possibility of more severe biological problems, could change such balances for this and other plantation species.

Eucalyptus deglupta, first planted on a commercial scale at Jari in 1979 and occupying almost 20,000 ha by 1983, has grown unevenly and has proved to be sensitive to drought. Another species, *Eucalyptus camaldulensis*, has performed better in these respects than *E. deglupta*. Commercial planting of *E. camaldulensis* began in 1982 and reached 8,065 ha by 1983. The future performance of *E. deglupta* and *E. camaldulensis* will have to be monitored closely.

Planners in Brazil have suggested that the project at Jari be emulated on a large scale in other parts of the Amazon Basin. Paulo de Tarso Alvim (1978c), an influential voice in Amazon development planning, believes that the results obtained at Jari "clearly demonstrate the enormous potential for commercial silviculture of the Amazon." Others have suggested that Jari represents an "experiment" that will provide a development model once it proves itself by making a profit. Ample reason exists to doubt the wisdom of applying Jari as a model for large-scale developments in other parts of the Amazon (Fearnside and Rankin 1980, 1982a,b, 1985).

Forest Exploitation

Forest exploitation currently is the subject of intense debate in Brazil. The Renewable Natural Resources Department of The Superintendency for Development of the Amazon (SUDAM) has proposed that "agricultural colonization" in Amazonia be replaced by "forestry colonization" in development plans. "The development of forestry colonization should be attempted through the creation of Income Forests *[Florestas de Rendimento]*, which would offer a chance for population nucleii to spring up in the Amazonian interior with a tendency toward stable and significant growth from the socioeconomic point of view" (Pandolfo 1978:66). These Income Forests, a set of twelve areas totaling 39,504,000 ha (Pandolfo 1978:22), or 7.9 percent of Brazil's Legal Amazon, would be the responsibility of a government enterprise, which would determine extraction and management techniques to be applied, and supervise "the contracts of exploitation concessions" (Pandolfo 1978:50). Reforestation after cutting would be done by the government, with the cost paid by the logging enterprises (Pandolfo 1978:63).

Forest exploitation became an issue of public controversy in Brazil when more concrete plans were laid for the granting of exploitation concessions. In December 1978 a report submitted to the Brazilian government by an FAO forestry expert based on a two-week visit to Brazil suggested that "forest utilization contracts" be instituted to grant private firms logging concessions on government land in Amazonia: "This form of forest concession basically constitutes a risk contract between the government and the private sector" (Schmithüsen 1978:13). Some press reports claim the plan would cover 56 million hectares (11.2 percent of the Legal Amazon or 20 percent of the "dense" forests) (Frota Neto 1978), while others give figures as high as 40 percent of the "Amazonian forest" (A *Crítica* [Manaus], December 23, 1978, p. 3). No area figures are given in the original report (Schmithüsen 1978). Clara Pandolfo, the chief proponent of the SUDAM "Income Forest" scheme, has denied there is "any relation or similarity between the Income Forests . . . and the so-called risk contracts" (Pandolfo 1979:2).

While head of the now dissolved PRODEPEF, Mauro Silva Reis called for establishing "conservation units for the purposes of multiple use," including the "rational logging and production of wood for industrial use" (Reis 1978:19). The proposal states that "the liberation of these areas, however, should occur only after Brazil has the technology and know how needed to rationally manage the heterogeneous tropical forest of Amazonia" (Reis 1978:12). It is also cautioned that "in truth, a self-sustained system of production for the dense tropical forest for industrial ends, based on the model considered here, has not yet been developed" (Reis 1978:14).

Following the public outcry in Brazil sparked by announcement of the forest utilization contracts ("risk contracts") scheme, all decisions were postponed pending a complete revision of Amazonian forest policy. A special interministerial commission was formed to produce a new draft law in a period of only 120 days, which ended on October 10, 1979. Subsequent drafts of the proposed law removed many restrictions on deforestation and ranching included in the commission's original version. The types of colonization to be promoted in the region, and policies concerning fiscal incentives for ranching, forestry, and other development schemes, as well as many other related issues, are all included under the rubric of "forestry policy."

CHAPTER TWO
The Tropical Rainforest As an Ecosystem

AN ECOLOGICAL SYSTEM is an identifiable set of interacting biotic (living) and abiotic (nonliving) components. The tropical rainforest is a particularly fragile and complex ecological system, or ecosystem, made up of highly diverse species of plants and animals. The extent to which this environment can support human settlement requires an understanding of its ecological features.

Ecological Features

Vegetation

The vegetation of Amazonia is divided into several distinctive types, each with specific agricultural problems and potentials (table 2.1). The most common forest type in the Brazilian Amazon is the forest of *terra firme*, or nonflooded high ground.

Excluding the *cerrado* (scrub savanna) zone of the Central Brazilian Plateau, *terra firme*, the nonflooded uplands, accounts for 70 percent of the 4.99 million km^2 total area of Brazil's Legal Amazon, or 94% of the 3.70 million km^2 that is phytogeographically Amazonian or "hylean." The recent drive for colonization has concentrated on the *terra firme* rainforests. *Várzea*, a periodically flooded mostly grassland vegetation type, has far higher agricultural potential than upland types because soil fertility in the *várzea* is renewed periodically by deposited silt. Most *várzea* is flooded annually by the seasonal rise in river water levels, although a small amount of *várzea* exists in coastal areas flooded daily by fresh water rising with ocean tides. *Várzea* occurs along silt-laden "white water" rivers such as the Solimões (Upper Amazon). *Igapó*, or swamp forest, is a flooded vegetation type that occurs along the nutrient-poor "black-water" rivers such as the Rio Negro. Unfortunately, maps and discussions of the Amazon often lump *igapó* with *várzea*, despite major differences in soil, vegetation, and agricultural potential.

Table 2.1. Major Ecological Zones of the Brazilian Amazon

Zone	Description	Vegetation Type	Area (km^2) [a]	Percent of Legal Amazon	Principal Locations	Reference
Terra firme	High ground	(total)	(3,487,000)	(69.88)		Pires 1973
		Dense forest	3,063,000[a]	61.39	Most of hylea	Braga 1979
		Liana forest	100,000	2.00	Xingú–Tapajós	Pires 1973
		Bamboo forest	85,000	1.70	Acre, Rondônia, SW Amazonas	Braga 1979
		Submontane forest (encosta)	10,000	0.20	NW edge of Amazonia	Pires 1973 Braga 1979
		Dry transitional forest	15,000	0.30	Southern edge of Amazonia, especially S. Pará	Pires 1973 Braga 1979
		Savanna (campo)	150,000	3.01	Roraima, Amapá, Humaitá, Marajó, Trombetas	Pires 1973
		Open white sand scrub (campina)	34,000	0.68	Rio Negro	Pires 1973
		Closed white sand forest (campinarana)	30,000	0.60	Rio Negro	Pires 1973
Várzea	Floodplain	(total)	(70,000)	(1.40)		
		Forest	55,000	1.10	River margins Amazon and tributaries especially above Parantins	Pires 1973

			Lower Amazon intergrades with *várzea* forest	Pires 1973
Igapó	Savanna	15,000	0.30	Pires 1973
Swamp	Swamp forest	15,000	0.30	Pires 1973
Coastal	(total)	(128,000)	(2.56)	
	Mangroves	1,000	0.02	Rio Negro and tributaries
				Pires 1973
				Braga 1979
	Dunes (*restinga*)	1,000	0.02	Pará, Amapá Pires 1973
				Braga 1979
Montane	Low montane (*serrana*)	26,000	0.52	Pará Pires 1973
Other types and water surfaces		100,000	2.00	Roraima Braga 1979 Pires 1973
Total Brazilian Hylea		(3,700,000)	(74.14)	
Parts of Legal Amazon not phytogeographically Amazonian	Dry scrub savanna and forest (*cerrado*)	(1,290,520)	(25.86)	Maranhão, Goiás, S. Pará, Mato Grosso	Pires 1973
Total Legal Amazon		(4,990,570)	(100.00)		de Almeida 1977

[a] Other estimates of dense forest area: 3,048,000 km² (Pires 1973:182); 2,800,000 km² (Reis 1978:4).
[b] The value for the area of the Legal Amazon that has been used in vegetation studies differs slightly from both the 4,975,527 km² value used by INPE in LANDSAT studies (Tardin et al. 1980:11) and the IBGE value of 5,005,425 km² (table 2.1). Part of the difference is accounted for by 27,138 km² of water surfaces (Brazil, IBGE 1982:28), and by a change in the southern boundary of the region (see chapter 1, note 1).

Soils

Soil quality directly affects any agricultural undertaking. The illusion deluding both early explorers and recent settlers that the large trees of the rainforest indicate exceptionally rich soil is quickly dispelled for anyone who sets out to farm on most soils in the Amazon. Almost all of the soils of the *terra firme* are of very old geological origin, and many essential plant nutrients have been leached out (Irion 1978; Van Wambeke 1978; Sombroek 1966; Bennema 1975; Falesi 1967, 1972a, b, 1974a; Camargo and Falesi 1975; Verdade 1974). With rare exceptions, the soils are extremely acid. Phosphorus is an especially scarce element in these soils, a fact aggravated by an unfortunate synergism with low pH, which further reduces phosphorus availability to plants. Overall, the major soil types in the Brazilian Amazon *terra firme*, specifically the area traversed by the Transamazon Highway between Estreito and Itaituba, are characterized by highly variable soil quality, usually with low nutrient content and low pH (table 2.2).

Not all *terra firme* soils are uniformly poor. On the Transamazon Highway several relatively limited patches of *terra roxa* (ALFISOL) exist west of Altamira. During the early days of colonization in that area, much emphasis was laid on the higher fertility of this soil. Falesi's assessment that "these soils can be cultivated continuously for over ten years, and the harvests always compensate the farmer" (1974b:2.7) is quite optimistic, since yields decline after a few years on *terra roxa* as on other soil types, and crop failures have occurred for a variety of reasons. *Terra roxa* does have a much more favorable pH than other soil types, often having values of 6.0 or more under virgin forest, as compared with pH values of 3.8–4.5 for the more common soil types. Phosphorus, as in other soil types, is unfortunately much lower than plant requirements. It is not uncommon to find total phosphorus levels in *terra roxa* less than one part per million, or one-tenth the level considered optimal for most crops. *Terra roxa* is nevertheless far better than other soil types in the Amazon *terra firme*, with the lone exception of tiny scattered patches of anthropogenic black soil on the sites of former Amerindian camps *(terra preta do indio)*, too limited to affect colonization significantly. Unfortunately, very little *terra roxa* exists in the Amazon, and estimates of the amount have been decreasing steadily as the Amazon has become better known.[1] Most recent figures (e.g., Falesi 1974b:2.8) estimate the area of *terra roxa* at only about one five-hundredth of the area of the Legal Amazon. The very limited extent of these relatively fertile soils makes agricultural exploitation of the vast Amazonian uplands a risky and unpromising enterprise.

Table 2.2. Major Soil Types of Transamazon Highway Colonization Areas

Soil Type (Brazilian System)	USDA Classi-fication[a]	% of transect[b]	Agricultural potential	pH (in H_2O)	Al^{+++} (meq 100g)	P_2O_5 (meq 100g)	K^+ (meq 100g)	Ca^{++} (meq 100g)	Mg^{++} (meq 100g)	Na^+ (meq 100g)	N (%)	C (%)	CEC[c] (meq 100g)	Sample depth (cm)
Terra roxa	ALFISOL	9.6	good	5.8	0.00	0.15	0.29	5.31	0.81	0.05	0.21	1.56	9.60	0–20
Yellow latosol	ULTISOL	18.2	poor	3.9	1.56	0.22	0.03	0.04	0.06	0.03	0.13	1.24	8.85	0–30
Red-yellow podzolic[d]	ULTISOL	38.6	poor	4.1	2.89	0.21	0.06	0.10	0.09	0.05	0.11	1.09	6.77	0–20
Lateritic concretionary	Petroferric Paleudult	6.9	very poor	4.5	2.25	0.45	0.26	0.45	0.98	0.06	0.13	1.25	8.81	0–20
Quartzose sands[e]	Quartzi-psamment	17.9	very poor	4.8	0.40	0.23	0.03	0.10	0.02	0.04	0.04	0.22	1.84	0–20
Others[f]	Others	8.8	poor–very poor	4.2	6.1	0.06	0.20	1.1	1.6	0.05	0.17	1.19	15.9	0–28

NOTES: These soils are from the 799 km of the 1,254 km section of the Transamazon Highway between Estreito and Itaituba for which soil identifications are reported by Falesi (1972a). The three Transamazon Highway colonization areas at Marabá, Altamira, and Itaituba are all within this area. The percentage of more fertile soil (*terra roxa*) shown here is substantially more than exists in the Estreito-Itaituba stretch as a whole, since none of the 455 km (36 percent of the total distance) not reported are *terra roxa*. The percentage of *terra roxa* for the Transamazon Highway as a whole is even lower, since none of the remainder of the highway (west of Itaituba) has *terra roxa* (Brazil, EMBRAPA-IPEAN 1974). The 76.8 km of *terra roxa* reported represents 2.6 percent of the approximately 3000 km of the Transamazon Highway in Amazonia.

The soil chemical information given in the table is for the superficial layers of typical profiles. Data are from the following sources: *terra roxa*, yellow latosol, red-yellow podzolic, lateritic concretionary, and quartzose sands are from Falesi (1972a:136, 69, 168, 99, and 108, respectively); data for "others" are from Brazil, DNPEA (1973a:57).

[a] U.S. D Agr. (1960); seventh approximation classification equivalents from Beinroth (1975); Sánchez (1976); and Brazil, RADAM (1978 18:271).
[b] Percentage of 799 km reported by Falesi (1972a). In some cases where more than one soil type were reported for a given highway segment, the stretch is apportioned equally between the types.
[c] Cation exchange capacity (sum of Ca^{++}, Mg^{++}, Na^+, K^+, H^+, and Al^{+++}).
[d] Side-looking airborne radar survey (Brazil, RADAM 1974, vol. 5) classifies a number of areas as yellow latosol, which are classed as red-yellow podzolic by Falesi (1972a).
[e] Most of the quartzose sands (142.4 km or 17.82 percent of the transect) are distrophic red and yellow sands. One small area (0.4 km, or 0.05 percent of the transect) is white sand REGOSOL.
[f] Brunizem 2.4 km, or 0.30 percent; cambisol (INCEPTISOL) 30.9 km, or 3.87 percent; grumosol (VERTISOL) 7.0 km, or 0.88 percent; slightly humid gley (tropaquept) 10.7 km, or 1.34 percent; hydromorphic soils 18.0 km, or 2.25 percent; alluvial soils, 1.0 km or 0.13 percent.

Plants and Nutrient Cycles

Tropical rainforest trees possess myriad features that minimize the limitations of poor soil fertility. The vast majority of nutrients in the ecosystem is tied up in the vegetation itself, rather than in the soil (Fittkau and Klinge 1973). Very tight nutrient-cycling mechanisms result in minimum leakage from these virtually closed systems. These mechanisms include direct nutrient cycling by plants in association with mycorrhizal fungae (Stark 1970, 1971, 1972; Went and Stark 1968). A fairly large proportion of the trees are legumes, many with nitrogen-fixing *Rhizobium* bacteria in root nodules. Soil fauna and litter communities are large and diverse, ensuring a rapid recycling of nutrients from dead leaves, branches, or animals. The forest is so efficient at trapping nutrients that in the Rio Negro Basin many nutrients are found in higher concentrations in the rainwater entering the system than in the streams leaving it (Herrera et al. 1978).

The closed nutrient cycling of the rainforest is in sharp contrast to most agricultural systems, which are designed for the export of nutrients in the products harvested. These systems generally also lose massive quantities of nutrients through leaching and erosion, especially with the torrential rain in the Amazon. The need for designing and promoting agroecosystems that minimize these losses is urgent.

Species Diversity

High species diversity, both in terms of number (species richness) and evenness of densities (species abundance), is characteristic of tropical rainforests generally (Richards 1964). Of great importance for agriculture is the protection this diversity affords against outbreaks of pests and diseases. The distance separating members of the same species in the natural forest helps insulate them from attack by pests, a fact that Janzen (1970a) offers as an explanation for this extraordinary diversity. (For other theories, see Pianka (1966, 1974).)

The reliance on monocultural agriculture in tropical rainforest areas carries special risks for this and other reasons (Janzen 1973a). The classic example is the rubber plantation at Fordlândia. The rubber trees, which survive low levels of disease attack in their native Amazonian forest, were attacked by a devastating outbreak of the South American Leaf Blight fungus *Microcylus ulei* (syn.: *Dothidella ulei*) when planted in a monoculture (Gonçalves 1970:11; Sioli 1973). The diversity of trees in tropical rainforest and the resulting dispersion of members of the same species inhibit the spread of specialized pests and diseases.

Diversity is particularly important in the tropics, where disease and pest problems are far greater than in temperate regions. The high humidity and temperature of the tropics, and the absence of a cold winter are ideal for many crop disease organisms (Janzen 1970b, 1973a). An extended dry season can have some of the pest-reducing effect of winter, but in no way brings tropical pest problems down to temperate zone levels.

Seasonality

Rainfall in most regions of Amazonia varies seasonally and among years (figure 2.1). During dry seasons, the soil in open agricultural fields can become as dry and parched as that in arid or semi-arid regions, limiting agricultural potential. The unpredictability of exactly when and how long each drought will last causes many crops to fail. In the case of bean crops on the Transamazon Highway timing of planting is a delicate problem—if planted too long before the rains taper off, the beans *(Phaseolus vulgaris)* will be attacked by a fungus *(Rhizoctonia microsclerotia* syn.: *Thanatephorus cucumeris)*.

Figure 2.1. Class one (poor) virgin burn. (Gleba 17, Lot 50, 1975).

Environmental Concerns of Rainforest Use

The use of the rainforest for agriculture, usually involving deforestation, radically alters ecological features such as vegetation, soils, nutrient cycles, and species diversity. Short-sighted patterns of rainforest use have aroused widely shared concerns. It is important to examine some of these concerns, including some popular misconceptions concerning environmental consequences, so that pioneer farming can be assessed in terms of its environmental cost and chances for long-term sustainability.

Highly diverse natural ecosystems, such as tropical rainforest, are strongly affected by disturbances. Compare, for example, the difficulty of a rainforest returning to its original composition following clearing and burning with that of a second growth stand of *Cecropia* in the same area. The second growth stand lacks the diversity and complex web of coevolved relationships of the mature rainforest. These differences, as well as their implications for fragility, can be missed by observers impressed by the fast formation of tropical second growth stands. Jari's former owner Daniel K. Ludwig, for example, is quoted in a rare interview as disputing the classification of rainforest as a "surprisingly fragile ecosystem" by observing: "Hell's bells, I spend five million dollars a year just to whack down the wild growth that springs up among our planted trees!" (McIntyre 1980:710).

Oxygen: A Straw Man

The purported threat to the world's oxygen supply from tropical deforestation is one of the more unfortunate misconceptions related to rainforest use, especially in Brazil. Oxygen levels are actually quite stable (Van Valen 1971) and are not dependent on rainforests, which use up as much oxygen as they produce (Farnworth and Golley 1973:83–84). The idea that the Amazon rainforest is responsible for the world's oxygen supply has gained particular force among the popular press in Brazil, where the Amazon is called the "lung of the world." This belief came into prominence after a popular Brazilian periodical interviewed Dr. Harald Sioli, and later misquoted this distinguished figure in Amazonian research (Sioli 1980). After exposing the oxygen argument as fallacious, it is usual to imply that all arguments linking deforestation with climatic change, including the important questions of carbon dioxide and rainfall, are "alarmist" and unworthy of serious attention.

Carbon Dioxide: "Greenhouse Effect"

Carbon dioxide (CO_2) plays an important role in the balance controlling global temperature. CO_2 is released when carbon stocks, such as forests and fossil fuels, are burned. Atmospheric carbon dioxide classically is considered the cause of a "greenhouse effect," where energy in the form of visible and ultraviolet rays from the sun passes through the atmosphere freely but is unable to escape when reradiated in the form of infrared radiation.[2] An increase in carbon dioxide would result in the earth's climate warming as more energy was trapped by the atmosphere. Atmospheric carbon dioxide increased linearly from 1850 to 1960, but has since been increasing exponentially. By 1978, CO_2 levels had only increased by 18 percent over the levels of 1850, but they are now expected to have doubled by sometime in the next century. Global temperature increases of 2° to 3°C have been predicted to result from this development (Stuiver 1978).

The amount of warming that would result from a doubling of atmospheric carbon dioxide is not known with certainty. A U.S. National Academy of Sciences expert committee has estimated an effect of $3°C \pm 1.5°C$ (United States NAS 1979, see Wade 1979).[3] The Academy estimated that current trends would lead to a doubling of 1979 CO_2 levels by 2030, with a few decades more needed for saturation of the heat-absorbing capacity of deep oceans before uncontrollable temperature rises take place (Wade 1979). Models of Manabe and Stouffer (1979), which include seasonal insolation fluctuation and a less idealized modeling of geography than earlier models (Manabe and Wetherald 1967, 1975) show a mean warming of 2°C, but with significant regional and seasonal asymmetries. Regional differences can have a great potential effect. Early estimates of a doubled effect at the poles (Budyko 1969) have been reinterpreted to yield a lower value of 25 percent above the mean global warming (Lian and Cess 1977).

Using the U.S. NAS estimate of $3°C \pm 1.5°C$, the possibility of a mean warming by even the lowest of these figures resulting in the melting of polar ice caps has concerned a number of meteorologists. The disproportionately higher temperature increases at the poles are especially worrisome.

> According to most of the recent research, the Arctic ice sheet can just maintain itself under present climatic conditions. Therefore, significant further warming would cause a complete transformation by the creation of an open sea in place of the Arctic ice sheet; an open Arctic Ocean should result in the drastic movement of all climatic zones several hundred kilometers northward . . . The effect of such a shift would be

especially noticeable in the belt which presently has a subtropical climate with winter rains (California, Mediterranean, Near East, and Punjab), which would then become arid steppe (Flohn 1974:103).

The drying of many parts of the globe would be accompanied by increased rainfall in some presently dry regions (Revelle 1982; Schware and Kellogg 1982).

Some uncertainty exists as to the rapidity and magnitude of the rise in sea levels that would result were polar ice to begin to melt. The contribution of Antarctic ice is particularly uncertain, as much of it is poorly mapped (Thomas et al. 1979). A 5 m rise in ocean levels is expected (U.S. CEQ/DS 1980 see Marshall 1981); early estimates of a 35 m potential rise (Goodland and Irwin 1975a:35) are probably high. The speed with which melting of the West Antarctic ice sheet would occur is a matter of uncertainty, estimates ranging from less than 100 years to several centuries (Mercer 1978; Thomas et al. 1979). Recent surveys of expert opinion favor a time scale of centuries for the breakup of Antarctic ice, with sea levels rising at an accelerated but less catastrophic rate of 70 cm/century (Kerr 1983).

Although more reliable and detailed data, especially from the tropics, are needed before firm conclusions can be drawn on the future of world temperatures, the simple doubt that major and irreparable meteorological changes could occur should give pause to planners intent on promoting massive deforestation.

DEBATE ON THE WORLD CARBON PROBLEM

Predicting future CO_2 levels and their effects is complicated by other climatic factors that could nullify some of the global warming, as they have done since 1940. One of several simulations for modeling global climate finds the net result of deforestation to be overall global cooling, mainly due to increased albedo, or reflectivity, of cleared land as compared with forest (Potter et al. 1975; see also Sagan et al. 1979). The rash of contradictory predictions should not obscure recognition of the delicate balances on which these processes depend, and the woeful lack of data on some of the most important parameters, especially in the tropics. In addition to lack of reliable data on deforestation rates, biomass, and nonliving carbon pools such as charcoal, climate models have shown themselves to be particularly sensitive to such poorly quantified parameters as atmospheric CO_2 levels before the industrial revolution (Björkström 1979:446-52) and the rate of mixing of the ocean layers serving as sinks for both carbon (Björkström 1979) and heat (Dickenson 1981:433).

Much of the carbon dioxide increase has historically resulted from burning fossil fuels. The biosphere has been singled out as a key factor by several studies (Bolin 1977; Hobbie et al. 1984; Houghton et al. 1983; Woodwell 1978; Woodwell et al. 1978, 1983). Marked seasonal oscillations in CO_2 levels, especially in temperate zones, testify to the importance of the biosphere in maintaining this delicate balance. Since tropical rainforests are estimated to contain 41.5 percent of the world's plant mass of carbon, and tropical seasonal forests another 14.1 percent (calculated from data of Whittaker and Likens 1973:358), the world carbon problem could be affected by the fate of tropical forests.

The incomplete burning of forest biomass, a substantial amount of which remains as charcoal, moderates the effect of forest burning (Crutzen et al. 1979). Lacking data from the tropics, Seiler and Crutzen (1980) used an estimate of unburned biomass based on observations following a wildfire in a temperate stand of ponderosa pine *(Pinus ponderosa)* to estimate the size of the world carbon sink in elemental carbon remaining in burned areas. This sink, estimated at 0.4–1.7 billion metric tons, together with estimates of the rate of deforestation lower than those used by other modelers, plus a substantial sink in afforestation, led Seiler and Crutzen (1980) to conclude that the land biota could be either losing or gaining 2 billion metric tons of carbon per year. This figure is much lower than the loss estimates of 4 to 8 billion metric tons per year calculated by Woodwell et al. (1978). The root cause of such sharp discrepancies is the rudimentary nature of data available, especially on tropical deforestation, forest biomass and carbon content, growth rates of tropical second growth, and burning efficiencies. Research to fill these gaps in knowledge should be a top priority, especially in the Amazon. A review of biomass studies in the Brazilian Amazon arrives at an estimate of 60 billion metric tons C in live and dead forest biomass (above and below ground) plus 5.1 billion metric tons C in the top 20 cm of the soil (Fearnside 1985a). Transformation of the area to pasture would release a total of 62 billion metric tons C as the vegetation biomass declines to the 2 metric tons/ha wet weight level characterizing Amazonian pastures (Hecht 1982:355) (=0.21 billion metric tons C in Brazil's 5×10^6 km^2 Legal Amazon at 0.475 dry matter content and 0.45 carbon content), and as the carbon in the top 20 cm of the region's soil falls to 3.14 billion metric tons (using 0.56 g/cm^3 forest soil density, 0.91% C content under forest and 0.56 percent C content under pasture).

Nitrous Oxide: Ozone Depletion

Rainforest clearing appears to be one of the contributors to a global increase in atmospheric nitrous oxide (N_2O). This gas is known to react in the stratosphere to produce nitric oxide (NO), which in turn serves as a catalyst in the breaking of ozone (O_3) molecules. Evidence for a strong catalytic effect comes from observations in nature (Fox et al. 1975), although rates for these reactions are quite low (Ruderman et al. 1976 note 6). Stratospheric ozone acts to absorb incoming ultraviolet radiation, shielding the biosphere from intense UV radiation.

The injection of N_2O into the stratosphere by proposed supersonic transport (SST) aircraft was a subject of heated debate during the mid 1970s. Ozone depletion effects of fluorocarbons from aerosol propellants and refrigerants became a public issue during the same period. Unfortunately, the potential ill-effects claimed were occasionally exaggerated, causing many to cease worrying about ozone depletion in subsequent years. Loss of public interest in stratospheric ozone was also partly the result of a widely publicized summary of a report by the US government's Climatic Impact Assessment Program (CIAP) which "conceals the logical conclusions of the study" (Donahue 1975). The understatement of effects identified during the course of the original study was later bitterly pointed out by the atmospheric scientists involved (see exchange of letters in *Science*, March 28, 1975, pp. 1145–46), but could not undo the effect on public perceptions stemming from wide press coverage of the CIAP report's "Executive Summary" (Grobecker et al. 1974). Even more unfortunately, the realities of nitrous oxide and ozone depletion are still with us and are likely to increase.

Increased UV radiation could be expected to increase substantially the incidence of skin cancer (basal cell carcinoma, squamous cell carcinoma, and melinoma) in humans: a 10–20 percent reduction in ozone could be expected for example to increase UV by 20–40 percent raising skin cancer incidence by about 20 percent among the Caucasian population of the world (Donahue 1975). More recent estimates double the effect of ozone depletion on skin cancer, each 1 percent depletion leading to a 2–5 percent rise in skin cancer average over the U.S. population (United States NAS 1982b, cited by Maugh 1982). Any possible behavioral changes in UV oriented insects should be determined by actual testing before making claims. Effects on aquatic ecosystems are numerous and deserve close scrutiny due to the key role of aquatic organisms in many food chains and biogeochemical cycles (Calkins 1982).

Possible effects in agriculture due to increased rates of mutation cannot be predicted with confidence with available knowledge, but the disastrous consequence of negative impact on any of the staple grain species is ample cause for avoiding exposure. DNA's absorption maximum is at 260 nanometers, only slightly below the 286 nanometer present lower limit of solar radiation reaching the earth's surface (Eigner 1975:17). One of the principal concerns is the expected deleterious effect on agriculture from increased mutation rates.

The impact of rainforest burning on nitrous oxide flux to the atmosphere, as well as the seriousness of expected changes, are areas of current debate. The debate illustrates both the minimal level of our present understanding of many fundamental global processes and the near total absence of relevant data, especially from the tropics. The concentration of N_2O in the troposphere has been increasing at about 0.2 percent per year (0.5 ppbv/year) over the past twenty years (Weiss 1981). All known sources of N_2O are at ground level, and many are linked to human activities. One source is the decomposition of organic materials in low oxygen environments, such as much human waste deposited in the anoxic conditions of dumping sites or sewage water (McElroy et al. 1976). In addition to wastes and compost, agriculture produces nitrous oxide through aerobic nitrification of fertilizer nitrogen (Bremner and Blackmer 1978). Combustion of fossil fuel is a major source, believed to account for about half of the total 1.1×10^{11} moles N_2O annual anthropogenic input (Weiss 1981; Weiss and Craig 1976). Nitrous oxide production from deforestation is believed to be significant from two sources: combustion of the felled biomass (Crutzen et al. 1979) and increased production in bare soil as compared to forest (Goreau 1981). Forest soils have been found to produce significant fluxes of N_2O through oxidation of ammonia by nitrifying bacteria, with rates increasing at low oxygen levels (Goreau et al. 1980; Goreau 1981). Cleared land, however, produces much more N_2O than does the same area under forest cover.

The contribution of fertilizer to global N_2O flux needs to be better understood as a check on the share attributed to deforestation. The importance of oxygen concentration gradients in nitrifying environments has recently been demonstrated by Goreau (1981). Much of the N_2O produced through denitrification at deeper (less well oxygenated) layers in the soil is never released to the atmosphere, but rather is consumed within the soil as an electron acceptor in respiration reactions (Goreau 1981:78). Much of the work done with fertilized agricultural soils has not taken this uptake into account (T. J. Goreau, pers. comm. 1982). The implication of this is that estimates of N_2O

production in fertilized soils probably exaggerate the N_2O derived from fertilizers—and a larger share of the observed atmospheric increases must therefore be explained by other sources, such as deforestation.

Nitrous oxide flux measurements from the tropics are nonexistent. Several indirect indications, however, suggest the conclusion that deforestation in tropical forests results in larger N_2O fluxes than temperate equivalents. Low counts of nitrifying bacteria are characteristic of acid soils under tropical forests (Nye and Greenland 1960; Jordan et al. 1979), but the nitrifiers greatly increase in numbers when clearing and burning raises soil pH (Nye and Greenland 1960). When the humus, root mat, and detritus are oxidized in the exposed soil, increased nitrification would release corresponding amounts of N_2O.

The long-term contributions of rainforest felling are unclear. One reason is the large amount of rainforest converted to cattle pasture. Initial conversion to pastures would result in release of N_2O as with all clearing. The lower equilibrium organic matter content of soils under pastures as compared with tropical forests (see Fearnside 1980c) would contribute to this, as soil nitrogen is approximately 98 percent organic (Russell 1973). Grasslands are known for low nitrification rates (Nye and Greenland 1960; Russell 1973), which would mean that further releases of N_2O from the soil should be relatively small once the initial conversion had taken place. Nitrous oxide would continue to be released from combustion, however. Pasture is burned frequently while it lasts, and after being invaded by second growth can be expected to undergo cutting and burning at intervals of a few years until weeds, compaction, and soil fertility degradation force abandonment of stock raising. Savannas are often burned as a matter of cultural tradition in Brazil, even when no immediate economic use is intended.

Hydrologic Cycle: Desertification

The issue of "desertification" is an emotional one, especially in Brazil with reference to the Amazon. A tendency toward decrease in rainfall in the region, even if not crossing the threshold of annual precipitation that defines a desert in climatological terms, is a possible consequence of deforestation which cannot be dismissed (Fearnside 1979c, 1985b). One reason is that in the Amazon, far more than in other parts of the earth, rainfall is derived from water recycled into the atmosphere through evapotranspiration, rather than being blown into the region directly as clouds from the Atlantic Ocean.

Estimates for the contribution of evapotranspiration to the precipitation in the Amazon Basin as a whole range from 54 percent based on an estimated annual total precipitation of 12.0×10^{12} m^3 and river discharge of 5.5×10^{12} m^3 (Villa Nova et al. 1976) to 56 percent based on water and energy balances derived from average charts of wind and humidity (Molion 1975). More detailed studies of the area between Belém and Manaus have produced estimates of the evapotranspiration component in rainfall in this part of the Basin ranging from 48 percent, based on calculations of precipitable water and water vapor flux (Marques et al. 1977), to up to 50 percent (depending on the month), based on isotope ratioing (Salati et al. 1978). Western parts of Amazonia, such as Rondônia and Acre, depend on evapotranspiration for a greater portion of their rainfall than does the Belém–Manaus area where estimates were made, and therefore would be expected to suffer greater decreases when forest is felled.

Hydrological work near Manaus has shown that a mean of 66 percent of evapotranspiration is transpiration rather than evaporation (calculated from Leopoldo et al. 1982). Both evaporation and transpiration are positively related to leaf area. Clearly the much greater leaf area of rainforest as compared with pasture, crops or second growth indicates deforestation will lead to decreased evapotranspiration and consequently decreased rainfall in the region.

Reduced evaporation and precipitation within the Amazon Basin would reduce the source of water vapor for neighboring regions. Rainfall in Brazil's Central-South Region—the country's agricultural breadbasket—would be jeopardized (Salati and Vose 1984).

Other consequences of deforestation, such as increased albedo (the ratio of reflected to incident light) also affect rainfall. Some models predict decreases in rainfall in temperate regions as a result of increased albedo provoked by tropical deforestation, leading to lowered heat absorption, reduced evapotranspiration and heat flux, weakened global air circulation patterns and reduced rainfall in the 45°–85°N and 40°–60°S latitude ranges (Potter et al. 1975). However, the magnitude of changes in albedo resulting from deforestation is a matter of debate. Simulation work by Henderson-Sellers and Gornitz (1984) indicates negligible effect of possible albedo changes in the Amazon on the climate of other parts of the globe, although the simulations show "severe" albedo-induced decreases in precipitation and evaporation within the region. Problems arise from differing definitions of albedo, and from use of unrealistic values for forest albedos prior to clearing. Forest vegetation reflects only a small amount of visible light, as indicated by its dark appearance. However, a large amount of reflectance occurs in the near infrared region of the spectrum, making forest albedos much higher if infrared radiation is included in the measurement. Dickenson (1981) has criticized studies such as that of Potter et al. (1975) for using visible spectrum albedo values (Posey and Clapp 1964) derived from measurements made in the temperate zone between 1919 and 1947 (List 1958: 442–43). Widening the spectrum included in albedo measurements and using a suitable average of more recent values from the tropics approximately doubles the albedo of forest from 0.07 to between 0.12 and 0.14 (Dickenson 1981:421). Combined with the assumption that forest is replaced by green secondary vegetation, albedos of these areas increase by only 0.02 to 0.04, or one half to one fourth the increases assumed by Potter et al. (1975) and others. The assumption is critical that evergreen vegetation replaces primary forest, however, as open savanna or grassland resulting from decreased rainfall in dry periods (e.g., Salati et al. 1978) and repeated burnings by humans (Budowski 1956) could well be a more likely future for these areas.

The illusion must be dispelled that, because annual rainfall totals in the Amazon are quite high, a significant amount of drying could be tolerated. The dry season in the Amazon already poses severe limits on many agricultural activities. During the dry season of 1979, Manaus went for 73 days without a single drop of rain. Soil water levels fell very low both in the open and under forest cover, where trees continued to transpire from large leaf areas. Since plants react to water levels in their root zones on a day-to-day basis, and not to the abstraction of annual rainfall statistics, effects of even small increases in the severity of the dry season could be dramatic. Rainforest, which does not tolerate severe water stress, could be expected to gradually give way to more xerophytic *cerrado* (scrubland) vegetation. Such a change would have the potential for becoming a positive feedback process, where the resulting further reduction in evapotranspiration would increase dryness and accelerate vegetational changes (table 2.3).

Genetic Diversity: Extinction of Species and Ecosystems

The genetic diversity of Amazon rainforest is legend. One inventoried hectare 30 km from Manaus had 235 species of woody plants over 5 cm in diameter (Prance et al. 1976). Many of the Amazon's species of plants and animals have never been collected or described, and each new collecting expedition reveals several new species (Prance 1975; Pires and Prance 1977). Many Amazonian species are endemic to the region and occur in limited ranges. Thus extensive deforestation would automatically ensure extinction

Table 2.3. Possible Macro-Climatic Effects of Amazonian Deforestation

Item	Change	Effect
Oxygen	Not significant	Not significant
Carbon dioxide	Increase	Global temperature increase (note: contribution of rainforest is subject of controversy)
Nitrous oxide	Increase	Global temperatures increase (slightly); ultraviolet radiation increases at ground level
Albedo (reflectivity)	Increase	Decreased rainfall in temperate zones
Evapotranspiration	Decrease	Decreased rainfall in Amazon and neighboring regions; temperature increase due to decrease of heat absorbing function of evapotranspiration
Rainfall	1. Decrease in total 2. Increased length of dry season (more important)	Vegetation changes: climatic regime becomes unfavorable for rainforest; reinforces trend toward still dryer climate

SOURCE: Fearnside 1985b.

of many species. The potential loss of genetic diversity from deforestation in the Amazon has been a major concern of biologists worldwide (Eckholm 1978; Ehrlich 1982; Ehrlich and Ehrlich 1981; Lovejoy 1973; Myers 1976, 1979, 1980a, b, 1984; Oldfield 1981). Genetic diversity offers potential for discovery of new organisms of economic value, or new uses for already known organisms—for example, new crop plants and varieties. The continuing evolution and dispersal to new areas of crop pests and diseases means that need for new germplasm will never cease. A good example is the vital input of coffee germplasm from the remnants of forest in Ethiopia as a means of obtaining resistance to leaf rust *(Hemileia vasatrix)* in *Coffea arabica* (Oldfield 1981). Destruction of stands of disease-tolerant, if low-yielding, natural rubber trees in Acre and Rondônia is one of many such losses occurring due to Amazonian deforestation. The same applies to need for new pharmaceutical chemicals in the face of continuing evolution of human disease organisms. The rush to obtain natural quinine when malarial parasites evolved resistance to chloroquine is a case in point (Oldfield 1981). The value of rainforest as a resource for fundamental scientific research has also been argued (Budowski 1976; Jacobs 1980; Poore 1976).

Ecological diversity, as well as genetic diversity in the strict sense, is quickly destroyed by deforestation. Often complex coevolved associations become extinct long before the last individuals of the species involved disappear (Janzen 1972a, 1974, 1976).

Indigenous Groups: Disappearance of Tribal Peoples

When the Transamazon Highway was announced as a way to bring "people without land to a land without people" the statement was tragically in error. Virtually all of the Amazon was already occupied when the highway construction program was launched. The large areas not settled by Portuguese-speaking "Luso-Brazilians" were occupied by Amerindians (Davis 1977). The incompatibility of colonization with the maintenance of indigenous populations in these areas is obvious. Locations of Amerindian tribes with relation to proposed highway routes are described in Robert Goodland and Howard Irwin's book *Amazon Jungle: Green Hell to Red Desert?* (1975a, ch. 5; not included in the Portuguese language edition, Goodland and Irwin 1975b). The resolution of conflicts of interest between highway construction and indigenous populations has rarely been nondestructive for the Amerindians (Davis 1977; Bodard 1972; Ramos 1980; Hanbury-Tenison 1973; Brooks et al. 1973; Bourne 1978; de Oliveira et al. 1979).

Most observers agree that since indigenous cultures are not compatible with "development," the solution is to separate Amerindian groups from settlement areas through provision of adequately sized, located, and protected reserves. It is a question of bitter debate as to where reserves should be placed, how large they should be and whether reserves should be respected when land is desired for highway routes, mineral deposits, ranching, agriculture or land speculation.

The tropical rainforest is regarded as a resource for pioneer farming by the Brazilian government, as well as by the thousands of individuals and groups that have set out to replace rainforest with agriculture in the Amazon. Characteristics of the rainforest ecosystem, changes that occur after it is cleared and planted, and environmental and other considerations tied to the massive scale of these alterations all must be considered in planning colonization programs and other forms of development.

Effects of Agricultural Operations on Soil

ANNUAL CROPS

Annual crops are planted as a part of "slash-and-burn" agriculture in traditional systems of shifting cultivation where a fallow period restores soil fertility and other characteristics needed for crop production, allowing future re-use of the same plot. Even though colonists engaging in pioneer farming usually do not have the cultural adaptations which go with this traditional type of shifting, or swidden, cultivation, they burn and plant in much the same way at least for the first step of the cycle. Later, colonists may deviate from the shifting pattern by planting perennial crops or pasture, by selling the lot to someone else who plants these things, or by allowing the land to lie fallow but not for the time period required to maintain production on a long-term basis. Nevertheless, much of what is known of shifting cultivation and its effects on soils applies to colonists engaged in pioneer farming. This knowledge has been reviewed in the classic work of Nye and Greenland (1960), as well as in such more recent reviews as those of Sánchez (1973, 1976), Andreae (1974), Watters (1971), and Greenland and Herrera (1978).

BURNING

Great variability in burn quality stems from the unpredictability of weather, individual variation among colonists, and delays in securing bank loans for clearing. Poor burns, a major problem in the colonization area, have a dramatic effect on crop yields. In addition to the problem of unburned vegetation, if a burn is poor (figure 2.2), the soil does not receive the beneficial increase in the levels of pH, phosphorus, calcium, magnesium and other cations, or the equally beneficial decrease in the level of toxic aluminum ions.

All fields prepared from virgin forest or second growth (at least 8 months uncultivated) are burned before planting ($N = 247$ virgin + 54 second growth), while weeds (fallow more than 2 months and less than 8 months) are only burned 27.1 percent of the time ($N = 48$).

FALLOWING

Fallowing helps to restore the potential agricultural productivity of the soil. Carbon and nitrogen levels are little affected by normal burning, since the stores of these elements in the vegetation are volatilized during combustion. If the burn is too hot, it is possible to lose these elements from the soil as well. These elements increase in the soil under second growth (Greenland and Nye 1959), and decrease after clearing both from cropping and from simple exposure of the soil (Cunningham 1963; Nye and Greenland 1964). Soil structure also recovers under fallow from compaction suffered during cropping. One study in southern Brazil (Freise 1934, 1939, cited by Budowski 1956:26) found an increase in pore volume from 12 percent in manmade savanna to 38 percent under 17-year-old second growth, or a 78 percent recovery in relation to the 51 percent value for pore volume under virgin forest in the area. Recovery from soil compaction under short periods of bush fallow is slow: Popenoe found that mean soil bulk density (which is inversely proportional to pore volume) in four sites in Guatemala increased from 0.56 g/cc under forest to 0.71 g/cc on cleared land under shifting cultivation, but only returned to 0.70 g/cc under two-year-old second growth (1960:122).

A period of fallow followed by a burn also reduces the populations of pests and weeds. The proximal cause for a tropical farmer's abandoning a given field in favor of shifting to a new location is generally the declining crop yields in relation to investment in labor, especially labor during critical periods limiting total production. Farmers do not care about the levels of phosphorus or any other element in the soil, being concerned only with the ultimate result of their efforts in terms of yield. Yield decline is the cumulative effect of a wide array of factors, of which individual soil deficiencies are only a part. Increase in weed competition with successive years of continuous cultivation can often be at least as important as soil depletion (Popenoe 1960; Greenland and Herrera 1978). Due to greater ease of measurement, much of the research done to date on shifting cultivation concentrates attention on soil changes rather than on pests, weeds, or diseases. Variability from these biological factors is included in the models for carrying capacity estimation on the Transamazon Highway, although soil effects were the focus of the greater part of the data collection effort. In view of the importance of these aspects of tropical agroecosystems, the current state of knowledge is woefully inadequate.

Restoring soil quality through fallowing in shifting cultivation is a cheap and effective strategy where population density and land prices are low. Usually, traditional shifting cultivation systems are characterized by fallow periods of lengths varying greatly from place to place, often as long as 20 to 30 years. One definition of "shifting cultivation" characterizes it as having 10 percent or less of the total cycle spent under cultivation (Ruthenberg 1971). In many long fallow systems, fallow can be shortened up to a point without impairing soil fertility (Guillemin 1956). Arguments have been advanced for using shifting cultivation as a basis for agroecosystems designed to support tropical populations at higher densities and living standards (Clarke 1976, 1978). Low energy inputs and high energy efficiency (not considering the energy released in burning fields) are advantages. Compare energetic input/output ratios of 1:16 for shifting cultivation (Rappaport 1971) with 1:3 (Pimentel et al. 1973) to 1:1 (Stout 1974, cited by Greenland and Herrera 1978) for U.S. agriculture. Shifting cultivation provides relative stability and nondestruction of large areas of forest *if* the population is maintained at a low density; a major weakness is that this stability and nondestructiveness are lost if population pressure does increase. The dominant view among development planners in Brazil is that shifting cultivation should be eradicated as quickly as possible (Alvim 1977a, b, 1978a, b, c). Chief disadvantages are generally seen as destruction of timber and increased erosion (United Nations, FAO 1959), and the limited potential for producing an exportable surplus, with the hoped-for possibility of increasing the cash standard of living (Alvim 1978a, b, c). Planners favor such replacements as annual crops with fertilizers and other chemical inputs, various perennial crops, pasture, silviculture, and forestry management schemes.

Pasture

The question of soil fertility under pasture has been of more than academic importance in Brazil, since claims that pasture improves soil fertility have lent support to government incentives programs. The idea that pasture improves soil quality has been expressed repeatedly at scientific meetings throughout Latin America since 1973. Falesi (1974b:2.14) compared soils under virgin forest and pasture of various ages on the Belém-Brasília Highway at Paragominas in Pará and at the Suiá-Missu Ranch in northern Mato Grosso:

> Immediately after burning [of forest] the acidity is neutralized, with a change in pH from four to over six and aluminum disappearing. This situation persists in the various ages of pastures, with the oldest pasture being 15 years old, located in Paragominas. Nutrients such as calcium, magnesium, and potassium rise in the chemical composition of the soil, and remain stable over the years. Nitrogen falls immediately after the burn but in a few years returns to a level similar to that existing under

primitive forest. . . . The formation of pastures on latosols and podzolics of low fertility is a rational and economic manner in which to occupy and increase the value of these extensive areas. (Falesi 1974b:2.14–2.15)

The belief that pasture improves soil carries over into official suggestions for land use, with the recommendation that extremely poor soils be planted as pastures. EMBRAPA's recommendations for the Transamazon Highway west of Itaituba call repeatedly for "the formation of pastures which, when well managed, cover the surface of the soils completely, protecting them from erosion, at the same time reinstating the biological equilibrium" (Brazil, EMBRAPA-IPEAN 1974:43). Even an hydromorphic laterite described as "tending irreversibly to hardpan" is recommended for pasture (Brazil, EMBRAPA-IPEAN 1974:46).

The soil differences noted by Falesi (1974b, 1976) do not lead to the conclusion that pastures will be sustainable (Fearnside 1980c; Hecht 1981). High pasture grass yields cannot be sustained if growth is being restricted by low quantities of certain nutrients, such as phosphorus, regardless of the quantities of other nutrients. Using data from pasture fertilization experiments in Belém (Serrão et al. 1971), lack of phosphorus has been shown to limit pasture grass growth (see Fearnside 1979a). Low phosphorus has also been found to limit grass growth in Paragominas, leading to more recent government recommendations of phosphate fertilization (Serrão et al. 1978, 1979; Toledo and Serrão 1982). The data from Falesi's Belém-Brasília Highway study (1974b, 1976) show a strong downward trend in available phosphorus after an initial peak from the burning of the virgin forest. Available phosphorus (P_2O_5) falls from a high of 4.18 mg/100g in new pasture to a lower plateau after five years. The five-year-old pasture has 0.46 mg/100g of P_2O_5, and after some slight variation, the value is still 0.46 mg/100g in the tenth year (Falesi 1976:42–43), lower than the virgin forest soil value of 0.69 mg/100g.

Much of the debate regarding soil changes under pasture is irrelevant to the question of maintaining pasture productivity. The important question is: are the low values to which phosphorus levels fall under pasture adequate to sustain production? The answer is no, as poor yields confirm both in experimental and commercial plantings wherever fertilizers are not applied. Weed invasion reinforces this effect. Weeds make up a greater proportion of the total biomass in pastures with lower soil fertility, as shown by results in fertilization experiments (Serrão et al. 1971:19). Weed competition is a major problem that, combined with the effects of declining soil phosphorus, restricts pasture productivity. Soil compaction also contributes to pasture yield decline (Dantas 1979; Schubart et al. 1976). In experiments with the most common pasture grass species, Guinea grass *(colonião : Panicum maximum)*, second-year grass yields in Belém were 63 percent of first-year yields, and third-year yields were only 49 percent of first-year yields (Simão Neto et al. 1973:9).

FERTILIZATION

Maintenance of soil fertility through fertilization is central to many plans for farming in the Amazon. However, colonists rarely use fertilizer despite a series of government measures designed to encourage application of these inputs (Fearnside 1980b).

In the case of pasture, for which phosphate fertilization is now recommended and subsidized for large ranchers (Serrão and Falesi 1977; Serrão et al. 1979), the question of whether production can be sustained through fertilization is first one of economics, and ultimately one of nonrenewable resource stocks. The relative costs of fertilizers and the resulting production are basic to assessing the prospects for any crop. Fertilizers cost two to three times as much in the Amazon as compared with southern Brazil (Alvim 1973), and more fertilizer is needed to achieve the same result due to rapid conversion (fixation) of the phosphate applied to unusable compounds. For some nutrients the high rainfall also leaches much more of the fertilizer out of reach of plant root systems before it can be absorbed. In the case of phosphates it is fixation, rather than leaching, which is the principal problem.

On the Transamazon Highway, as in most of the Amazon, phosphorus fixation capacity of the soil is very high, with the greatest effects occurring at low (and most probable) levels of fertilizer application (Dynia et al. 1977). In red-yellow podzolic (ULTISOL) from the Transamazon Highway, up to 40 percent of the phosphorus applied is fixed after seven days, while *terra roxa* (ALFISOL) fixes up to 83 percent, at 75 and 53 ppm P applied respectively (Dynia et al. 1977). Such problems can be overcome by applying more fertilizer to saturate the soil's fixation capacity. Marked increases in pasture grass yields have been obtained in experimental plots on the Belém-Brasília Highway (Koster et al. 1977; Serrão et al. 1978, 1979) using 50 kg of P_2O_5 per hectare, which translates into about 300 kg/ha of superphosphate fertilizer. Similar experiments in Amazonian Peru have shown increases in productivity for a few years followed by a decline due to compaction and probable other nutrient deficiencies in addition to disease attack in associated legumes (Peru, IVITA 1976, cited by Sánchez 1977:563).

Erosion

Erosion is another problem that plagues agriculture in the Amazon. Many people not familiar with the region harbor the illusion that the Amazon is flat, an impression encouraged by the appearance of the forest from the air. Although some parts of the basin are indeed quite level, much of it is dissected into steep slopes. Erosion causes significant soil losses when the soil is exposed for cultivation, with soil surface often dropping one or two centimeters per year under annual crops (Fearnside 1980d; see also McGregor 1980; Scott 1975, 1978). Erosion has a detrimental effect on soil fertility, since the soil quality on the Transamazon Highway is generally worse at lower depths than at the surface (soil profiles in Brazil, IPEAN 1967; Falesi 1972a; Brazil, DNPEA 1973a, b; Brazil, RADAMBRASIL 1974, vol. 5). This effect contrasts with the situation in some other parts of the world where erosion can improve soil quality by exposing less-weathered material (Pendleton 1956; Sánchez and Buol 1975).

Erosion would be likely to constrain agricultural production most quickly in systems which leave soil exposed repeatedly. One such system is a proposed technology for obtaining continuous production of annual crops (Nicholaides et al. 1982; Sánchez et al. 1982; Valverde and Bandy 1982). Erosion is one of a number of potential problems making widespread use of the system difficult (see Fearnside n.d.[b]). The "flat ULTISOL" of Peru's Yurimaguas experiment station (Nicholaides et al. 1982) differs from much of Amazonia, especially the areas undergoing intensive colonization in Brazil. A land use survey of the Amazon basin indicating that half of the region has slopes of less than 8 percent (Cochrane and Sánchez 1982:151) is often cited by proponents of continuous cultivation of annuals (e.g., Sánchez et al. 1982).

The 50 percent figure is deceptive, however, due to the large scale maps used to classify topography and other contraints. For example, in a 23,600 ha area on the Transamazon Highway where a detailed slope map was made based on field measurements at 225 points, 49.3 percent of the tract was found to have a slope of 10 percent or more (Fearnside 1978:437, 1984e). The entire area was classified as less than 8 percent slope by Cochrane and Sánchez (1982:149). Soil erosion could therefore be expected to affect the long-term sustainability of any system that leaves the soil exposed to rain.

Laterization

Laterization, or more properly the formation of plinthites, has been an overstated danger in many popular accounts of agricultural problems in the

Amazon. The idea that this hardened material, largely composed of iron oxides, covers much of the tropics originated in early reports by nineteenth century scientists visiting the tropics (Sánchez 1976:52–54). More recently, fears that vast areas of the Amazon would turn to pavements of brick upon clearing have echoed through the popular press. The prospect of hardening of plinthite over large areas of rainforest with deforestation is remote (Bennema 1975). Areas where plinthite formation is a potential problem are now thought to cover less than 7 percent of the tropics as a whole (Sánchez and Buol 1975) and 4 percent of the Amazon (Cochrane and Sánchez 1982); recommendations have occasionally been made that such soils be cleared in the Brazilian Amazon (e.g., Brazil, EMBRAPA-IPEAN 1974:46). It is important that the problem of laterite not be dismissed as an overreaction to the exaggerations of the past.

Effects of different agricultural operations are summarized in table 2.4.

Table 2.4. Summary of Effects of Agricultural Operations on Soil

	Soil Changes[a]							
	pH	Al^{+++}	P	N	C	K	Ca^{++} & Mg^{++}	CEC	Compaction
Burning	+	−	+	~0	~0	+	+	+	+
Annual crops	−*	+*	−*	−*	−*	−	−	−*	+*
Pasture	+	−	−*	−	−	+	+	+	+*
Fertilizing, liming, and manuring	+	−	+	+	+	+	+	+	−
Fallow in second growth	+/−	+/−	+/−	+	+	+	+	+	−
Fallow + burn	+	−	+	+	+	+	+	+	−

[a]Changes with time under the treatment: + = increase, − = decrease, +/− = either increase or decrease depending on initial level, ~0 = little change (can be either + or −)

* = soil change effects believed to be most detrimental in the Transamazon Highway area. Included in modeling: pH, aluminum, phosphorus, nitrogen, and carbon. Not included: potassium, calcium and magnesium, cation exchange capacity, and compaction.

CHAPTER THREE
Population Growth and Carrying Capacity

Human Population in Ecological Communities

HUMANS DERIVE virtually all of their sustenance from ecological communities, exploited either in their natural state or modified into agroecosystems—ecosystems that include crops and other organisms used to supply human needs. Humans become a part of these ecosystems, both as a significant component in the systems' food chains and as the agent of intervention to increase energy and material flows in parts of the systems directly useful to people. As with all systems, some elements and pathways in these ecosystems are extremely sensitive to such human interventions, while others are highly resistant.

Most human agricultural activities set back natural trends of ecological succession to an earlier stage, for example, the replacement of forest with grasses. The earlier stages of succession have the advantages of a greater difference between gross primary productivity (the energy fixed in carbon compounds by plants in photosynthesis) and community respiration (energy released by plants and animals), leaving greater yields for humans. The herbaceous plants characteristic of these early phases also have the advantages of converting a greater proportion of available energy into the seeds humans consume and of yielding their produce with a minimum of delay, usually in less than one year. Crop breeding programs (also easier with annual crops) have greatly increased the production for humans.

Maintaining ecosystems in a preclimax state, that is, one in which plants and animals are not yet able to maintain their populations in a stable equilibrium over a long time period, requires a constant input of energy to reverse the trend to dominance of species less productive to humans. Manual weeding is the most direct, and also the most laborious, method. Periodic burning of second growth or overgrown pasture requires much less direct human energy input when succession has proceeded further. The combustion itself releases a great deal of energy, as does burning primary forest during the initial conversion to agricultural uses.

Input to agricultural operations in the form of fossil fuels, the major form of energy subsidy in temperate regions, is increasing in the tropics through "green revolution" crop and management practices (AAAS 1975). Altering energy flows within agricultural systems through these subsidies has been a principal strategy for increasing outputs to humans: it has also resulted in great dependence on the continued availability of the energy subsidy (H. T. Odum 1971, 1983).

Nutrient cycles in agroecosystems are generally more open than those of the climax communities they replace. Some nutrients "leak" from the system due to the more limited recycling mechanisms, while others are removed through the produce harvested. When nutrients leach beyond the reach of the shallow root systems of crop plants, they can to a certain extent be returned to the surface by the deeper roots of woody second growth species during a fallow period. The faster rate of increase in second growth biomass during early succession makes it most efficient to cut stands on fallow sites before regrowth begins to slow (Ahn 1979). Nutrient stores in the vegetation are therefore never fully replenished. Nutrient stocks become increasingly depleted as the cropping and fallowing cycle continues, leading to slower regrowth. Fallow intervals must lengthen to reach the requisite size for slashing and burning. Concentration of major nutrient stores in the vegetation in tropical forest and second-growth makes these systems particularly vulnerable to disruption of nutrient balances. Cutting and burning vegetation is almost always a part of agricultural use in tropical rainforest areas, in contrast with temperate ecosystems where the major nutrient stores in the soil are rarely purposefully disturbed in conversion to agriculture.

Many ecological interactions of unmodified ecosystems are manipulated in agroecosystems. The density-dependent response of predator populations to prey availability, for example, can control outbreaks of pests. This kind of biological control is sometimes purposefully introduced as a means of restoring key feeding relationships limiting fluctuations in the more diverse natural ecosystems.

The problem of pest control in agricultural systems is accentuated by the fact that humans have bred crop species to remove most natural chemical and other defense mechanisms (Janzen 1972c, 1973a). The usual response has been to replace the plants' natural chemical defenses with applications of pesticides. Pesticides, in turn, cause a series of ecological effects in the treated ecosystems, as well as in ecosystems to which the more persistent compounds are eventually transferred. Small farmers in the New World tropics use very little chemical pesticide, although new cash crops and extension efforts are promoting their increasing use.

Changes in human density affect many of the basic relationships among ecosystem components and the relative abundance of component populations. As with all populations, human populations are subject to limiting effects at high densities, which affect consumption levels and environmental quality in many ways. Understanding population dynamics and their relationship to density-dependent effects is essential to an estimation of carrying capacity.

Population Growth Patterns

Exponential Growth

In his *Essay on the Principle of Population* (1798), Thomas Malthus recognized that, given abundant resources, human populations have an innate tendency to increase geometrically, or exponentially, producing a J-shaped growth curve (figure 3.1a).

The growth pattern can be described by

$$N_t = N_o e^{rt}$$

where:

N_o = number of individuals at time o
N_t = number of individuals at time t
e = base of natural logarithms (the constant 2.71828 . . . .)
r = innate capacity for increase given the prevailing environmental conditions
t = time

The mental translation of a small yearly percentage increase to a total population size at some future date is difficult. This is greatly facilitated by expressing growth rates in terms of the time required for the population to double (table 3.1).

Logistic Growth

Exponential growth at any rate greater than zero theoretically yields infinite population sizes if continued unchecked. Real populations only follow the exponential pattern when resources are abundant and other conditions are favorable for continued growth (Malthus 1798, 1830).

Verhulst (1838) derived a logistic equation to describe the sigmoid S-shaped

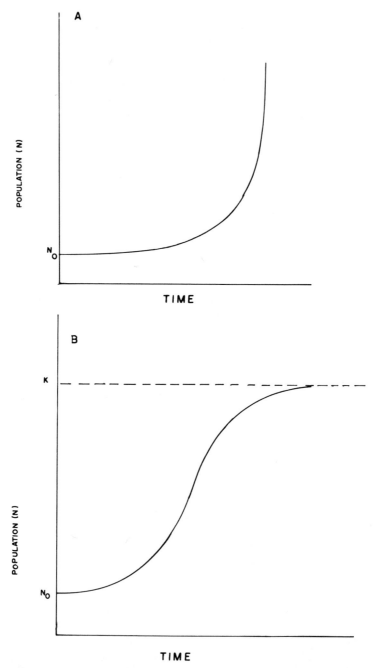

Figure 3.1. A: Exponential growth curve, where numbers increase rapidly toward infinity. B: Logistic growth curve, where initial exponential growth is slowed by density dependent effects.

growth pattern exhibited by a population with limited resources. At low density the population grows in the uninhibited exponential fashion, with growth rates tapering off at high densities as resources become insufficient to sustain continued population growth. In time the population asymptotically approaches an upper limit known as K, the saturation density or logistic carrying capacity (figure 3.1b). The difference between the exponential pattern and the logistic pattern in the curve is the effect of "environmental resistance," or what Malthus called the "difficulty of subsistence."

The logistic equation is usually written as a differential equation, indicating change in the *rate* of increase:

$$\frac{dN}{dt} = r N \frac{(K-N)}{K}$$

where:

dN/dt = rate of population increase
N = population size
r = innate capacity for increase given the particular environmental conditions (= the instantaneous birth rate minus the instantaneous death rate, for population closed to immigration and emigration)
K = saturation density or logistic carrying capacity.

In integrated form, the equation yields:

$$N = \frac{K}{1 + e^{a-rt}}$$

Table 3.1. Population Doubling Times for Selected Areas

Area	Year	Yearly Increase (%)	Doubling Time (years)	Reference
U.S.	1974	0.8	86.6	Ehrlich et al. 1977:96
World	≈1976	1.8	38.5	Ehrlich et al. 1977:96
Brazil	1960–70	2.85	24.4	Brazil, IBGE 1982:74
Brazil	1970–80	2.34	29.6	Brazil, IBGE 1982:70
Amazon[a]	1960–70[a]	3.4	20.3	Brazil, IBGE 1982:74
Amazon[a]	1970–80	4.9	14.1	Brazil, IBGE 1982:74
Manaus (city)	1970–80	7.1	9.8	Brazil, IBGE 1982:111
Rondônia	1960–70	4.6	14.9	Brazil, IBGE 1982:74
Rondônia	1970–80	14.8	4.7	Brazil, IBGE 1982:74

NOTE: Table assumes constant growth rates.
[a] Northern Region, i.e., Pará, Amapá, Roraima, Amazonas, Acre, and Rondônia.

where:

a = constant of integration defining the position of the curve relative to the origin (the value of ln $(K-N)/N$ when $t=0$)
e = base of natural logarithms (the constant 2.71828. . .)
t = time

The equation for logistic growth rests on a number of assumptions not met by many real-world populations of organisms: 1) an initial stable age distribution exists; 2) either all individuals are ecological equivalents or an appropriate density unit is used to account for differences in life history stage, individual size, etc.; 3) the innate rate of increase (*r*) is one that can actually be reached under the prevailing conditions, 4) there are no time lags; and 5) the relationship between density and rate of increase is linear, including the assumption of highest growth rate when population density is extremely low (Krebs 1972:194–95; Pielou 1969; Wilbur 1972; Wilson and Bossert 1971:104). Discrepancies between these assumptions and characteristics of real populations explain the many departures from the logistic growth curve of figure 3.1b found in nature. They also mean that much of the value of the logistic equation is as a description of growth under certain circumstances rather than as a predictor of growth where these conditions do not apply.

In the case of human populations, factors influencing birth and death rates are much more complex than in other species. This complexity is due to the greater control culture has permitted humans over both births and deaths, and over the environmental restrictions to which these rates must ultimately conform. Over the past two centuries of human history some of the cultural modifications of demographic behavior form a pattern: the demographic transition.

DEMOGRAPHIC TRANSITION

Demographic transition—the shift from a pattern of high birth and death rates to low birth and death rates—is seen by many as a great hope for third world countries avoiding the dismal "misery and vice" seen by Malthus as an inevitable consequence of the "difficulty of subsistence." Since World War II, most developing areas have been experiencing dramatic population growth as a result of lowering death rates through medical and public health improvements. Many of these countries have experienced some decrease of birth rates in more recent years, but all have birth rates substantially above death rates (Coale 1983). For the 1970–1980 period, Brazil had a crude birth rate of 33.0 per 1000 inhabitants, and a corresponding death rate of 8.1 (Brazil, IBGE 1982:108). Birth rates in Brazil, especially in urbanized areas, have been dropping in recent years. Preliminary results of the 1980 census indicate a drop in the mean number of children per family from 8 to 6 in the northeast (*Veja*, November 12, 1980, p. 28). The lowering of birth rates following decreases in death rates in developing countries bears many similarities to the historical demographic transition in developed countries, but differences in interpreting the significance of this similarity for development policies are sharp and impassioned.

The underlying causal rationale of the demographic transition theory is believed to be valid at a high level of generalization, in that correlates of development are associated with lowered fertility both within and among countries. In Latin America, those countries showing signs of beginning the lowered fertility phase of the demographic transition are the ones with most development (Beaver 1975, cited by Teitelbaum 1975).

Close examination of the demographic transition in nineteenth-century Europe and of recent trends in developing countries suggests that third world countries should not rely on these transitions to relieve rapidly increasing imbalances between population and resources (Teitelbaum 1975). Much greater and more rapid mortality declines in developing countries as compared with nineteenth-century Europe, together with higher initial fertility and the lack of opportunities for relieving pressures through international migration, mean populations are increasing at much greater rates, doubling in about one-third the number of years that characterized European populations during their fastest growing mid-transition phase. Developing countries could be caught in a vicious cycle, where effects of the rapid population increase subvert the fertility-reducing effects of the demographic transition: high densities and large family sizes result in lowered living standards, lowered educational opportunities, and less employment, especially for women, who are the first to lose in competition for scarce job opportunities (Newland 1977). These limitations, particularly as they affect women, are among the social factors believed to have the greatest effect on reproductive behavior (McCabe and Rosenzweig 1976; Ridker 1976).

For population densities below the point where fertility is restrained by poor nutrition, an unstable balance theoretically exists between two opposing feedback loops. Higher living standards result in lower fertility, leading to still higher standards; whereas any lowering of living standards results in higher fertility, causing standards to fall to still lower levels. Because of this second feedback loop, leaders of developing nations in the fertility-decline portion of the demographic transition would be well advised not to count on European history repeating itself in their countries. It is unlikely that the amount of economic progress realistically possible for many developing countries would be sufficient for the full fertility-reducing effect of the demographic transition to be realized. Even if realized, such an effect would be too slow-acting to prevent astronomical increases in population densities and intensification of problems following from these increases.

Fortunately, changes observed in some undeveloped parts of both nineteenth-century Europe and the modern third world suggest that reproductive behavior can be modified without development (Teitelbaum 1975). Development-associated effects of the traditional transition theory are also overshadowed by the effects of evenness in the distribution of wealth and social services such as education (Brown 1980:154–56; Ehrlich et al. 1977:782; Flegg 1979; Simon 1976). Small amounts of development are associated with much greater fertility declines if the benefits are evenly spread throughout society.

Development issues surrounding the interpretation of transition mechanisms are heated. The mechanisms themselves are not as simple as once was believed.

> There is great political virtue in a simplistic interpretation of transition theory asserting that in all circumstances development will "take care of" population matters. Nonetheless, scientists are obligated to report that close examination of transition theory in both historical and modern perspective shows that policy-makers would be ill-advised to adopt such a simplistic and deterministic view. (Teitelbaum 1975:178)

Factors Limiting Population Growth

Density-Independent and -Dependent Controls

Populations of all species are limited below the infinite levels to which exponential growth theoretically would lead them. Population growth can be limited by factors that are linked with the density of the population (density-dependent controls) or by physical factors such as unfavorable weather, flooding, or other natural catastrophes acting as density independent controls. Classification of controls as density-dependent or independent is frequently inappropriate, since density can often intensify the effects of "independent" physical factors. Exponential population growth, when limited primarily by a density-independent control, would be sharply truncated, dropping to a lower level to begin the exponential J-shaped climb again.

Density-dependent controls act through such mechanisms as competition between members of the same species for scarce resources (intraspecific competition), competition with members of other species (interspecific competition), or by predators, parasites, or disease organisms whose populations and/or attack frequencies increase in response to increased density of the prey or host population. Density-dependent controls usually act more gradually in limiting population growth than do density-independent ones, resulting in a tapering off of growth as an upper limit is approached (see figure 3.1b). However, density-dependent controls may come into effect after passing a threshold density, causing population levels to drop abruptly rather than declining smoothly to equilibrium. Oscillations produced in this way, and by time delays in density-dependent effects, are more undulating than the jagged boom-and-crash pattern of a population limited by density-independent forces. Often the oscillations tend to damp if environmental conditions are fairly stable.

Density-dependent mechanisms are more important in ecosystems with relatively stable physical conditions, such as the tropical rainforest, while ecosystems with violent seasonal changes (as in arctic or temperate habitats) have more populations limited by physical factors. Either form of limitation may operate by increasing the death rate, by reducing fertility, or both.

Checks on population growth act through the organism's requirements for survival and reproduction. When the supply of one of these requirements is insufficient to sustain population growth, the requirement is said to be a *limiting factor*. Lack of such requirements as light, water, nesting sites, and various nutrients may be limiting. Different factors may be limiting at different times. If an abundant source of a limiting factor is supplied to a population, the population will expand until supplies of some other factor are exhausted.

The common assumption that only one limiting factor restricts the growth of a population at one time is an oversimplification, since limitation of the population is the cumulative effect of the deaths and lost chances for reproduction by the individual members of the population, events that invariably arise from a number of different causes. Even the growth of a single individual is not always limited by a single factor, since supplying additional quantities of more than one factor can result in at least some increase in growth. The generalization of single limiting factors is close enough to the truth to be useful in such applications as fertilizer recommendations in agriculture. In other situations, the existence of multiple limiting factors has significant implications for the estimation of carrying capacity.

Definitions of Carrying Capacity

The term *carrying capacity* has been used by workers in biology, anthropology, geography, range management, fisheries, wildlife management, and business management with related but different meanings. All refer to the number of individuals that can be supported in a given area; the level of consumption at which they are to be supported and the time the area is to be capable of providing this support vary with the definition.

Carrying capacity can be classified by the time horizon of the estimate, yielding two categories: *instantaneous* and *sustainable*. The definitions can be further broken down according to whether they are static or dynamic; deterministic or stochastic; based on a single limiting factor, several possible limiting factors, or a combined measure representing the contributions of several factors. In static systems the values of all variables are constant through time, while dynamic estimates allow for changes with time. Deterministic estimates are based on fixed values for all parameters, while stochastic estimates include random variation in at least some of the parameters (with the result that the probability of an outcome is less than one). Since the real world is characterized by both changes with time and variability, dynamic stochastic estimates should lead to the most realistic estimates of carrying capacity.

Instantaneous Carrying Capacities

The logistic equation, describing the S-shaped or sigmoid population growth curve, is the forerunner of other classes of carrying capacity. Although I would not even casually suggest that this equation should be applied to humans, it has historically played a prominent role in human carrying capacity estimation. The logistic equation was derived by Verhulst (1838) and independently by Pearl and Reed (1920), and was intended in both cases for use with human populations. Pearl and Reed (1920) used the logistic equation to describe population growth in the United States based on census data collected every ten years since 1790. Revisions of the estimates were made after each census through 1940 (Pearl et al. 1940; Reed 1936). The results were so accurate that population could be calculated within a few thousand individuals, a feat that amazes statisticians (Snedecor and Cochran 1967:449). The shortcomings of the logistic equation as a tool for estimating human carrying capacities became more apparent in the years after 1940. Fitting the curve through the census of 1940 yielded a value for the saturation density or logistic carrying capacity (K) of 184 million for the United States (Pearl et al. 1940). That the population has increased far beyond this value should be less surprising than the remarkable fit obtained through 1940 using an equation as simple as the logistic. The logistic equation is based on a long list of assumptions, straining its applicability for organisms as simple as the cladoceran water flea *Daphnia* (Frank 1957), or even for a protozoan like *Paramecium* (Hairston et al. 1969). Kingsland (1982) has traced social processes within the scientific community that help explain the logistic's widespread use in the decades following 1920, even in the face of contrary indications. Despite the equation's severe limitations, the practice of calculating logistic carrying capacity for human populations by applying curve-fitting techniques to historical population data still persists (e.g., Schacht 1980).

Humans clearly do not conform to assumptions such as absence of age structure and time delays, or complete ecological equivalence of all individuals, to say nothing of interposing such a complex network of relationships as that represented by human culture between the "cause" of a given increase in population density and the "effect" of a given change in population growth rate. The term of the logistic equation representing "environmental resistance" bears little functional relation to processes actually at work in the United States during the period when initial exponential population growth began to slow in a demographic transition.

One way of modifying the logistic equation to alleviate some of its limitations is the addition of stochastic terms, done by Levins (1969) and May (1973:122). Computer simulations of populations of hypothetical organisms have shown that high variability in logistic carrying capacity leads to higher extinction rates and to lower population sizes (Roff 1974:264–65). Other continuous models exist that avoid some, but by no means all, of the logistic's restrictions.

The "carrying capacity" in the logistic equation (including its modified versions) sets the upper bound for the growth curve and is an instantaneous value related to the population's ability to survive and reproduce at given levels of resource consumption, not to the long-term sustainability of those levels of resource supply. Discussions of the logistic equation, especially in the context of application to human populations, often mistakingly assume that its carrying capacity value is sustainable (e.g., Hardesty 1977:195, otherwise a very useful review of human carrying capacity).

Examples of instantaneous carrying capacity calculations abound in the field of rangeland management. Most rangeland management usage of the term *carrying capacity* refers to a sustainable carrying capacity. However, some of the literature, including most of the Brazilian contributions to the field, is clearly using the term to refer to an instantaneous relationship, which this text will refer to as "short-term feeding capacity" when discussing pasture management and yields.

Some human carrying capacity estimation techniques determine when carrying capacity has been exceeded by some behavioral change in the population. Such behavioral changes indicate that the rate of production being obtained is unsatisfactory by the population's own culturally defined standard. These methods work only for populations observed during the period when the instantaneous carrying capacity is exceeded, or when separate subpopulations can be observed at the same time displaying differing behaviors at different densities. Examples include studies by Hunter (1966) in Ghana where emigration from densely populated areas indicated by changing sex ratios showed that this point had been passed, and by Vermeer (1970) in Nigeria where a shortening of the fallow period among shifting cultivators at high population densities indicated that the instantaneous carrying capacity had been reached. In the latter study, some broad indications can be deduced related to a sustainable carrying capacity as well, to the extent that the ten-year minimum fallow period traditionally in use in the sparsely populated areas appears to be sustainable, whereas the two-year fallow in the densely populated areas results in visible environmental degradation.

The information provided by instantaneous carrying capacity estimates such as these, when coupled with information from other studies concerning changes in soils, yields, and vegetation under different fallowing regimes, can lead to useful conclusions about sustainable population levels, with appropriate assumptions about technology and consumption. The principal problem with applying such methods is the need for comparable populations at different population densities ranging from levels below to levels above instantaneous carrying capacity.

Many of the shifting cultivation studies that have been done with the avowed intention of producing sustainable carrying capacity estimates would be more accurately categorized as instantaneous. The shifting cultivation formulas used in this type of study will be discussed in greater detail in the next section. Calculations are made without concern for sustainability in a broad class of permanent field agriculture studies as well (e.g., Cook 1970).

Sustainable Carrying Capacities

The basic definition of sustainable carrying capacity, patterned after that used by Allan (1949, 1965) in his pioneering work on estimating carrying capacities for shifting cultivators in Zambia (then Northern Rhodesia) is: *the maximum number of persons that can be supported in perpetuity on an area, with a given technology and set of consumptive habits, without causing environmental degradation.*

Archaeologists have made numbers of human carrying capacity estimates, usually basing the selection of a carrying capacity value on the observations that 1) the population being studied had successfully survived and reproduced at a given population density over a period of time, and 2) the population did not destroy soil or other resources in the process. The many ancient anthropogenic or human-made savannas throughout the tropics attest to frequent violation of the "harmony with nature" often assumed by investigators.[1]

Some anthropologists writing on contemporary and extinct aboriginal groups have used the same sorts of general observations on persistence and apparent equilibrium to draw qualitative inferences about carrying capacities (e.g., Meggers 1971). Many social behavior patterns have impacts on birth and death rates. Some authors switch between definitions of carrying capacity. An example is the "Club of Rome" modeling group, which summarized its computer simulations of world population, resource, and pollution trends in *The Limits to Growth* (D. H. Meadows et al. 1972). A logistic equation carrying capacity is used for part of the group's discussion (D. H. Meadows et al. 1972: pp. 100–1), but a sustainable carrying capacity is clearly the objective of the bulk of the group's writings, including the paper on carrying capacity of the globe (Randers and Meadows 1972). The most common reason for such confusion between instantaneous and sustainable carrying capacity is a failure to recognize the lack of connection between the levels of exploitation that correspond to maintaining survival and reproduction at any point in time and the rates of exploitation corresponding to avoidance of long-term degradation of the resource base.

SHIFTING CULTIVATION CARRYING CAPACITY FORMULAS

I. Allan 1949:14–15

$$\text{Area of land required per head} = 100 \, C \, L \, / \, P$$

where:

C = the cultivation factor, which is "an expression of the number of 'garden areas' required for each land type to allow the complete cycle of cultivation and regeneration normally practiced on that type under the system to which the calculation applies." A garden area is the "area in cultivation at any one time" (Allan 1965:30). This is calculated as:

$$C = \frac{\text{cultivation period} + \text{fallow per period}}{\text{cultivation period}}$$

where:

L = the mean acreage in cultivation at any one time per head of population
P = the cultivable percentage of the land type.

The "total carrying capacity" is the total land area available to the community divided by the area required per head.

II. Conklin 1959:63

Critical population size:

$$Cs = \frac{L}{A \, T}$$

where:

Cs = critical population size
L = maximum cultivable land available (conveniently expressed in hectares)
A = minimum average area required for clearing, per year, per individual (in hectares)
T = minimum average duration of a full agricultural cycle (in years).

Critical population density:

$$Cd = 100\ Cs\ /\ L$$

where:

Cd = critical population density (in people per square kilometer of L).

III. Carneiro 1960:230

$$P = \frac{T\ Y\ /\ (R + Y)}{A}$$

where:

P = the population of the community that "can be supported permanently in one locale"
T = the total area of arable land (in acres) that is within practicable walking distance of the village
Y = the number of years a plot of land continues to produce before it has to be abandoned
R = the number of years an abandoned plot must lie fallow before it can be recultivated
A = The area of cultivated land (in acres) required to provide the average individual with the amount of food that he ordinarily derives from cultivated plants per year.

IV. Gourou 1966:45; 1971:188

$$\text{Potential population density} = A\ C\ /\ B$$

where:

A = the number of cultivable hectares per square kilometer (= percent of the total area)
C = the number of inhabitants per hectare cleared each year
B = the length of the rotation (cultivation plus fallow).

V. Fearnside 1972:487–88

$$A = B\ D/C$$

where:

A = hectares per person at carrying capacity
B = average consumption/person/year

> C = yield of land of quality Q under agricultural system S/year
> D = number of unit areas needed for a long-term equilibrium migration cycle, where a unit area is the area of land which must be cultivated at any one time to support one person. The number of unit areas is given by:
>
> $$D = (E / F) + 1$$
>
> where:
>
> E = number of years required for abandoned land to recover
> F = number of years a plot can be farmed before abandoning.
>
> VI. Faechem 1973:234–35
>
> $$W = a / (C L)$$
>
> where:
>
> W = carrying capacity = maximum theoretical population
> a = cultivable area of land (ha)
> C = cultivation factor = number of garden areas required to complete a cycle of cultivation and regeneration = (fallow time + cultivation time)/cultivation time
> L = mean area presently cultivated per capita (ha/capita).

The Box formulas for calculating carrying capacity under systems of shifting cultivation can be reduced algebraically to a common form (Faechem 1973).[2] Faechem further reduces the result into an expression indicating that the ratio of what he calls the "theoretical population" to the current population is equal to the ratio of the land available to the land in use. An unstated assumption of Faechem is that the current population, as computed from the mean area per capita currently required to complete a full agricultural cycle, is in equilibrium. The assumption of equilibrium is indicated by farmed and fallow areas corresponding to the farmed and fallow times which are input as parameters in the formula. As Street (1969) has pointed out, the assumption of equilibrium is often unwittingly made by those who attempt carrying capacity estimates, resulting in circular arguments. Fortunately, the shifting cultivation carrying capacity formulas, despite a plethora of limiting assumptions, can still have some utility. If the equations' inputs are determined through measurements independent of other portions of the equations—determining fallow times based on studies of nutrient stores and land area requirements based on yield observations and nutritional requirements—then the information obtained from subsequent calculations is valid within the limitations of the assumptions on which the equations are based.

The "concept" of carrying capacity, as represented by the shifting cultivation formulas, has been attacked by Brush (1975) and Hayden (1975), and defended by Glassow (1978). Brush considers that "the principal empirical weakness of the concept of carrying capacity lies in the fact that the theory of homeostasis inherent to the concept is neither testable nor refutable" (1975:806). The "theory of homeostasis" here refers especially to a group's equilibrium-maintaining behavioral adjustments, which have been ascribed by investigators to changes in population density relative to carrying capacity. The key issue is the use that is made of carrying capacity estimates rather than the validity of the estimates themselves. When carrying capacity is used as an explanatory tool for observed changes in cultural patterns, plausible mechanisms must be identified by which the population's approach or passing of carrying capacity feeds back to the culture, both on the level of short term adjustments and on the level of long-term cultural evolutionary changes. For an entry into this debate see Brush (1976), Cowgill (1975), Vayda (1969, 1976), and Vayda and McCay (1975). The purpose of the present study of carrying capacity on the Transamazon Highway, however, is to provide an indicator that could be used in development and population planning rather than to explain demographic and technological changes.

Hayden believes that "the practical problems involved in measuring and using 'carrying capacity' have proven the concept to be deficient in theory, unrealistic in implementation, and impossible to measure" (1975:11). He proposes "abandoning" carrying capacity in favor of a measure called the *resource over-exploitation rate*. This rate is seen as a function of three variables: 1) the "potential resource intensity occurrence frequencies," which essentially is the frequency of dips in the availability of resources in the territory supporting the population, or "lean seasons" (Bartholomew and Birdsell 1953); 2) the technological potential; and 3) the population density. Hayden's measure has correctly given emphasis to the variability in supply of food and other resources, something lacking in the shifting cultivation formulas for carrying capacity. Hayden argues that the frequency, duration and severity of resource shortages (i.e., periods of exceeding the instantaneous carrying capacity) will be key factors affecting the response, if any, of a human population experiencing them.

Hayden's alternative measure is actually not a replacement for carrying capacity: if one solves the resource overexploitation rate equation for its population density term, and sets the resource over-exploitation rate at a maximum acceptable limit, one obtains a value very similar to carrying capacity as operationally defined for the present Transamazon Highway study, providing the additional criterion of sustainability were met. As with Brush's (1975) criticisms of carrying capacity, Hayden's main reservations concern the claims that have sometimes been made for shifting cultivation formula estimates as explanations of events in the evolution of cultures, rather than the problems of the shifting cultivation formulas themselves. Hayden's emphasis on the importance of variability is appropriate, not only to the archaeological interpretations he seeks, but also to the present study's task of developing carrying capacity as a planning tool.

Both Brush and Hayden despair of obtaining estimates of the parameters needed for carrying capacity calculation but, as Glassow (1978) points out, this is no reason for abandoning the attempt. The potential importance of carrying capacity in formulating sustainable population and development policies points to the need for much more effort, both in theoretical development and in data collection.

Bayliss-Smith (1980) has made a significant contribution to approaching carrying capacity in a way that could produce results usable by planners. Bayliss-Smith's method also allows estimation of what he calls "perceived carrying capacity," a quantity believed to be more relevant to the explanation of human behavior than is carrying capacity based on the capability of the environment to supply sustenance at a given level for an indefinite period. Bayliss-Smith's focus is on the relation of agricultural intensification to labor inputs and product outputs, the critical step being the construction of a graph of output per manhour versus output per hectare. Output per hectare is high at low values of output per manhour, but after a point, falls to a lower level as output per manhour increases. Carrying capacity corresponds to the point on the curve where output per hectare begins to fall sharply with rising output per manhour. The method goes beyond calculation of a single carrying capacity: it produces a matrix of values for the work input per productive person required to support a population not only at carrying capacity, but also at a series of consumption levels above the "subsistence" value used in defining carrying capacity. Such a matrix has clear value for planners contemplating the effect of different population policies on consumption and leisure.

Bayliss-Smith makes clear that his method gives emphasis to leisure time and surplus production, and leaves out factors such as variability in crop production and labor requirements (1980:62). His choices appear appropriate for the Fiji islands, the site of the UNESCO project of which his work forms a part. Significant differences between Fiji and the Transamazon Highway make other choices appropriate for the present study. The taro (*Colocasia* spp.) staple of the Fiji agriculturalists could be expected to produce relatively stable yields from year to year and from farmer to farmer, as is the case for most root crops in frost-free areas of the tropics. Variation in yields is a major concern for Transamazon colonists, whose upland rice is planted in slash-and-burn fields subject to poor burns and other hazards. In addition, the high value placed on leisure time by Fijiians is not shared by pioneer farmers in Amazonia, beyond observance of Sundays and a few other religious holidays. Most Transamazon farmers take great pains to appear busy at all times, and are often quick to apply scornful epithets to any of their fellows not visibly occupied. The present study of the Transamazon Highway therefore emphasizes variability in crop yields, and the many factors affecting these yields.

An Operational Definition of Carrying Capacity

Sustainable carrying capacity is operationally defined in terms of a gradient of probabilities of failure, or *failure probability–population density profile* (figure 3.2). Failure rates are those sustainable over some long time period at the corresponding human population densities. The criteria for failure can be defined in a variety of ways and can include multiple limiting factors or combinations of factors. They can include measures of environmental degradation as well as individual consumption. Focusing on individual consumption levels contrasts with the area-wide average consumption criteria implied by most definitions.[3]

The maximum acceptable probability of colonist failure, as well as the criteria for failure, can be chosen in accord with socially-defined values. Probability of failure increases with human density in an hypothetical relationship that should apply within some range of possible human densities. Note that the curve in figure 3.2 rises to a failure probability of one on meeting the vertical axis. The probability of failure would be expected to rise at low population densities due to a sort of "Allee effect," the phenomenon of reduced survival and reproduction at lower densities that is common to many species (see E. P. Odum 1971). For humans, the probability of failing to maintain adequate consumption standards would increase at very low densities due to the difficulties from lack of infrastructure, cooperaton, and other benefits of society.

Once a maximum acceptable probability of colonist failure has been se-

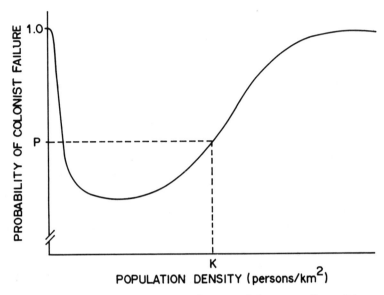

Figure 3.2. Carrying capacity (K) as determined from a gradient of increasing probability of colonist failure with increasing human population density (failure probability/population density profile).

SOURCE: Fearnside n.d.(e).

lected (point P in figure 3.2), the carrying capacity (K) is the corresponding population density. At greater densities, density-dependent effects cause the combined (density-dependent and -independent) probability of failure to exceed P. In a case where extremely high levels of risk cause the curve to exceed the maximum acceptable probability of colonist failure at all points, a reasonable solution would be to select the minimum probability of failure as the point corresponding to K.

Assumptions in Carrying Capacity Estimates

Assumptions of carrying capacity estimates often invalidate the techniques for their intended purposes. Street (1969) has identified several assumptions common in such studies and criticized such well-known workers in the field as Carneiro (1960), Conklin (1959), and Brookfield and Brown (1963) for lack of attention to assumptions. Street's most withering criticism is directed at their assuming that the farming and fallowing practices in use at the time of fieldwork do not result in environmental degradation. If degradation is taking place in an area, then the carrying capacity values obtained by substituting the observed fallow time, farmed time and so on for the parameters in a shifting cultivation formula will exceed sustainable carrying capacity.

A number of variables are often assumed to be constant over time in carrying capacity studies. These include technology, consumption patterns, and land use allocation (Street 1969). Other assumptions often unwittingly made in carrying capacity estimates, such as those based on shifting cultivation formulas, include the density-dependent effects of weeds and insect pests. Street points out that these biological problems can act to reduce yields as land use intensity increases. For example, the beneficial effect of a fallow in checking pest (Pool 1972) and weed (Popenoe 1960) populations is lost if fallow is eliminated in favor of continuous cultivation. In addition, studies based on independent evidence of soil quality degradation and regeneration rates must also face the inevitable problems of soil quality measurement, including the difficult problem of measuring "available" nutrients relevant to crop yield prediction.

VARIABILITY

A number of additional assumptions can be added to Street's list. Of prime importance is the assumption that has been the focus of the Transamazon Highway study: variability. The high levels of variability characterizing tropical agriculture will reduce carrying capacity both by necessitating planting a large buffer of additional land each year as insurance against poor yields and by reducing the margin protecting the population from failures due to both density-related causes and background levels of density-independent failures. In terms of the relationship depicted in figure 3.2, it is hypothesized that increasing variability would raise the curve upward in the region of relatively low probability of colonist failure at the left of the graph, including the point corresponding to the maximum acceptable probability of colonist failure (P). This would lower the carrying capacity from k_1 to k_2 (figure 3.3). The initial formulation did not anticipate the effect of variability in lowering the failure probability curve of figure 3.3 at very high population densities: this perfectly logical result was discovered later in running simulations of the Transamazon Highway system (chapter 5).

Many applications of the shifting cultivation formulas have assumed that a constant yield is obtained each year. One important exception is Allan's classic discussion of the "natural surplus of subsistence agriculture" from buffer

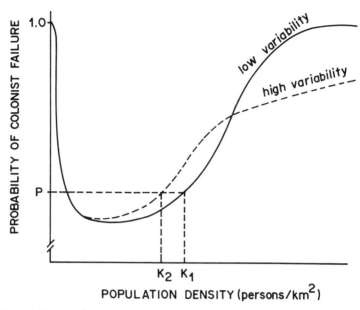

Figure 3.3. Hypothetical effect of variability on failure probabilities and carrying capacity.

SOURCE: Fearnside n.d.(e).

areas planted by subsistence cultivators in Zambia as protection against yield fluctuations (1965:38).

Limiting Factors

A recurrent problem in making carrying capacity estimates is that of selecting appropriate limiting factors. Most estimates using the shifting cultivation formulas are based on a single limiting factor, usually calories.[4] The same is true of the Forrester–Meadows world models. The choice of calories is often unfortunate, since the tropical populations for which most of the estimates have been made usually have far more readily available sources of calories from root crops than they do of protein, especially animal protein. Basing a carrying capacity estimate solely on calories can produce results at least an order of magnitude higher than estimates that include animal protein. Using calories from manioc alone, for example, de Fautereau (1952, cited by Carneiro 1960) estimated that 1,250 persons/km² could be supported in an area of French Guiana. The importance of protein has been recognized by Denevan (1970), Gross (1975), Lathrap (1968), and others, but carrying capacity estimates based on calories alone are still commonplace, e.g., Carol (1973) for tropical Africa and Shantzis and Behrens (1973) for the Maring of New Guinea, among many examples.

Howard T. Odum has struggled with the thorny problem of limiting factors in carrying capacity estimation, an ultimate goal of much of his work: "The essence of the problem of food production for the world is: what is the carrying capacity of the earth's surface for man?" (1971:125). At the same time he recognizes difficulties with his approach of converting all of the flows in the different systems modeled into kilocalories of energy: "The carrying capacity of an area may not be computed on the basis of gaining 3,000 calories of energy, for also required are the special components, each of which has an energy cost. The energy value of a vitamin is not its potential value as a fuel but the calorie expenditure required to manufacture it and deliver it to man" (124).

Howard T. Odum's solution to the problem of special components such as vitamins is to convert them into larger amounts of energy than would be gained from burning the same vitamin in a bomb calorimeter. The simplicity gained from converting everything into a common currency has been a valuable tool in promoting understanding of whole systems and of the parallels between different types of systems. It has also been valuable in elucidating the multitude of ways the present technological societies desperately depend on fossil fuel for their affluence and survival. Along with the advantages of being able to visualize an entire human system from a single page of flow-chart symbols are prices that must be paid for the loss of information. One is the magnification of errors from the conversion of tiny quantities, such as the vitamin in Odum's example, to large amounts of calories: small errors in the quantity of the vitamin would result in large errors in the caloric result. A more fundamental problem is the masking of limitations of resources that are not as easily substitutable as the caloric conversions would suggest. Culturally defined limits on acceptable substitutions can be an added complication (Bayliss-Smith 1974). A third drawback is one these analog models share with Forrester's system dynamics models for the digital computer (1970, 1971)—the loss of information about the nature and effects of variability in the different components, which results from condensing information representing an "average" value for an entire system into a single box, whether it be labeled "quality of land" or "kilocalories." All of these problems are manifestations of the difficulty of making carrying capacity determinations based on a single limiting factor.

The problem of limiting factors in carrying capacity has been addressed by Hubbell, who argues strongly against the "spate of single-factor answers in the last 20 years" (1973:95). He suggests instead that "several factors may act simultaneously, conceivably equally" in limiting the instantaneous carrying capacity. The same could be said for sustainable carrying capacity as well.

In his study of aboriginal populations associated with early Spanish missions in Baja California, Aschmann recognized that the problem of single limiting factors is intimately linked to that of variability:

> The seasonal availability of a particular food was probably of more significance than the amount present. The carrying capacity of the area, in terms of a human population which made little effort to store food, must be stated in terms of what was available in the poorest season of several years, not in terms of the average food supply. Consequently, a food available only in small quantity and ordinarily ignored may be the one that at critical moments prevented starvation. A consideration of only the ten or twenty most important foods may miss this critical aspect of the food economy. (1959:78)

Environmental Degradation

Closely associated with the problem of limiting factors in the context of sustainable carrying capacity estimation is the definition of environmental degradation. In studies focused on a single limiting factor such as calories, it is possible to circumvent this problem simply by equating degradation with anything that reduces the supply of the limiting nutrient and hence the carrying capacity. Carneiro (1960), for example, leaves degradation out of his definition entirely. A much more flexible treatment of sustainable carrying capacities is possible if restrictions on degradation can be added as additional limiting factors that allow an area to be viewed as a patchwork of differently classed subareas to which different standards of permissable degradation apply. Eugene Odum's paper (1969) on the strategy of ecosystem development points the way to this form of multiple criterion-based decision making on carrying capacities. The environment is viewed as a mosaic of patches allotted to different uses with different environmental standards to be maintained: some may be allotted to uses which result in a "degradation" by some criteria, while others may be required to remain in pristine condition. Maintenance of the integrity of the various kinds of forest and biological reserves in the Amazon set aside by the Brazilian government is an example of this type of criterion. Such criteria cannot easily be translated into common currencies such as kilocalories for use in single limiting factor models.

Selection of Appropriate Standards

Selecting appropriate standards for sustainable carrying capacities is not as easy as it may at first appear. Cultural biases of investigators of subsistence systems can often lead to inappropriate decisions. Nietschmann (1971) has reviewed the problem of bias in subsistence studies. However, in pioneer agriculture areas thoroughly integrated into the money economy, such as Brazil's Transamazon Highway, cultural bias does not pose quite such a difficult problem. Selecting criteria nevertheless remains a fundamental and somewhat arbitrary process.

Land Quality Classification

Assumptions related to land quality classification have posed difficulties in carrying capacity estimation, regardless of the technique applied. Some shifting cultivation formula studies have simply assumed constant quality of "arable" land (e.g., Carneiro 1960). Others have made adaptations of the basic formulas to accommodate different land quality classes.[5] In highly aggregated simulations such as the Forrester–Meadows world models, there is also a lumping of all "arable" land as equivalent. The generation of different soil qualities for individual patches of land in the KPROG2 model of the Transamazon Highway has reduced loss of variability from lumping land qualities into one or a few quality classes.

Economic Exchange

Assumptions regarding the isolation of the system under study from exchanges with the outside world can greatly affect carrying capacity results. For aboriginal tribes the assumption of isolation is often more or less warranted, but the situation can change radically when cash economies intrude. In Nietschmann's study of the Miskito Indians of Nicaragua (1972, 1974) this intrusion resulted in the destruction of the marine turtle populations on which the Miskito depended. Similarly, Gross and Underwood's 1971 study of northeastern Brazil indicated that farmers switching to a sisal-based cash economy could no longer buy the same quality of diet they had formerly enjoyed from subsistence agriculture. Variability in the market prices obtained for each crop can also make overdependence on exchange with the market economy a potential cause of colonist failure. H. T. Odum's 1971 discussion of the carrying capacity-elevating effects of outside power subsidies from fossil fuels represents the more widely appreciated effect of such exchanges.

Exchanges within the population, as well as those with the outside world, can affect carrying capacity by cushioning individuals against shortfalls and imbalances in their production (Freeman 1955; Sahlins 1972). Related to this is the inability to substitute specific food or other items for one another.

In the case of open systems such as that of the Transamazon Highway, exchanges with the outside relate closely to changes in farming technology and allocation patterns. Changes can result from new habits among the existing population accompanying importation of new seed varieties or other material or behavioral novelties or, as seems to be predominating at the moment, from the continuous flow-through of new colonist families with different behavior patterns.

Models for Estimating Carrying Capacity

Models constructed for estimating carrying capacity in the Transamazon Highway agroecosystem were designed to avoid the foregoing assumptions. The features of these models are discussed in chapter 5. It is hoped that the models will serve as a basis for future studies and will eventually suggest answers to some of the theoretical questions that underlie the estimation of human carrying capacity in tropical agroecosystems.

SIMULATION MODELS

The most practical approach to modeling complex ecological systems such as the Transamazon Highway region is the use of *simulations*. Simulations normally depend on mathematical calculations, or their equivalent, but unlike analytical models do not lead to a solution that has been "proven," or deduced from fundamental principles through a chain of mathematical reasoning. Rather, they use the relationships in the system, usually abstracted into a model as a series of equations, to calculate empirically what results are obtained with one or a series of sets of specific examples of initial conditions. Simulation enables the investigator to learn more about the real-world system that the model represents. Many models are designed specifically for use in simulation. Vast numbers of routine calculations are necessary in performing simulations with these models, since the effect of time is mimicked by "iterating," or repeating, the differential equations representing rates of change. Thus, the use of electronic computers is an asset in performing the simulations, both for their speed and the minimization of human error. Any simulation that can be done with the aid of a computer could equally well be done with a pencil and paper: all the computer contributes is speed.

Two types of computers are useful for simulations. The first is the analog computer, a form of physical model that substitutes electronic components such as resistors and capacitors for mathematical relationships. The components are wired together to represent the flows within the system, and a current (either constant or varying) is passed through the circuit. The voltage at different points tells the investigator how stocks and flows of energy or materials would behave in the analogous real system. Analog models, which are especially appropriate for representing the flow of energy in ecosystems, have been developed for a wide range of human, as well as nonhuman, systems (H. T. Odum 1971, 1983).

The second type of computer is the digital computer, which represents the system relationships by mathematical equations for numerical calculations. Digital computers are used for data analysis and an ever-increasing range of computational applications in everyday life, in addition to a wide variety of simulations. Their flexibility makes them highly appropriate for modeling the complex networks of relationships in real-world ecosystems.

Advantages of Simulation Models

Simulation models have a number of distinct advantages over others. Their results are quantitative and thus are less open to misinterpretations and easier to verify than qualitative statements. Because simulations are repeatable, chances are greatly improved for identifying and eliminating errors both as the initial model is developed and when the scientific community examines and criticizes it. Simulation also allows examination of hypothetical scenarios that would be impossible or impractical to test directly using the real-world system. Constructing simulation models imposes (or should impose) discipline on researchers: modelers must gather at least some data on all components of the model; none can be bypassed as often occurs in a prose description of an argument or model. Simulations can also integrate information about extremely complex systems, thus extending the human mind's reach in interpreting complicated phenomena. Conclusions reached are therefore less likely to be restricted to the realm of what has already been intuitively accepted by the investigator. The speed of simulations, combined with their repeatability, makes them ideal as frameworks for interpreting new information. Simulation models can be expanded to include additional items, or can be improved to make use of better information about items already included, much more quickly and reliably than other model types. All of these features make simulation models ideal for studying ecosystems, including agroecosystems.

Of course, researchers using simulation models and interpreting their results must be aware of certain pitfalls: oversimplification resulting from components selected for their ease of measurement rather than their importance to the real system's functioning;[6] failure to collect adequate data to parameterize the model, *i.e.*, obtain estimates of the model's inputs ("garbage in, garbage out"); and a tendency to extract information from a model that it was not designed to produce, e.g., deducing precise dates of future events from a model designed to reproduce system behavior modes.[7]

STEPS IN FORMULATING SIMULATION MODELS

The steps involved in conducting research with the aid of simulation models are very similar to those of the traditional scientific method (table 3.2), with the difference that the model must be designed and checked, and that experiments are performed on the model rather than on a part of the real-world system itself.

Gathering background information and identifying a problem do not necessarily occur in sequential order. Often a question is selected as important for theoretical reasons based on background information from one or more ecosystems (or even from theories unrelated to any actual ecosystems), and a search is then made to find a location appropriate for conducting the investigation. For example, in the study of carrying capacity of the Transamazon Highway, the effect of variability on carrying capacity was seen as important based on observations of agroecosystems in other countries, and only later as the Brazilian Amazon selected as the appropriate place to realize the study.

Once a problem has been defined and background information collected, ideas must be generated about how the system might operate. Such conjectures lead to a mental, or verbal, model serving as the starting point for more concrete testing of possibilities either through traditional experimentation or simulation.

"Brainstorming" to produce new and as yet untested theories is an essential part of the system operating to produce advancement of scientific knowl-

Table 3.2. The Two Methods

Traditional Scientific Method	Method with Simulation
Observation (background information about the system)	
Definition of the problem (question)	
→Generation of ideas ─────────→ (conjectures on how system functions = mental model)	Formulation of a formal model (identification of components and relationships, model purpose, and time horizon)
Formulation of an hypothesis (a testable statement)	Measurement of parameters
Identification of predictions following from the hypothesis	Quantification of functional relationships
Design of an experiment (manipulation of a part of the real-world system)	Simulation (experimenting with the model, including formulation of hypotheses and predictions and aquisition and analysis of data on results of model behavior)
Data collection	
Data analysis	Sensitivity tests (changed inputs or structure)
Interpretation of results (comparison with predictions)	Validation (comparison with known behavior of real-world system)
Conclusions about the real-world system ←─────	Conclusions about the model

edge in general. It plays a role analogous to generation of genetic variability through mutation and recombination in organic evolution. The "fundamental theorem of natural selection" (Fisher 1958) holds that the rate of evolution is proportional to genetic variability in a population. Much in the same way that natural selection acts to produce differential survival and reproduction among competing genotypes, the parallel process of evolution of science occurs through experimentation and comparison with rival theories.

Models have a number of *state variables*, or quantities representing the model components, affected by other components and by influences outside of the system boundary. Influences from outside the system that affect the components but are not affected by them are called *parameters* if constant and *forcing functions* if varying.[8]

Grouping real-world items into model components is an important step in formulating the model. Conventional wisdom among many (but not all) modelers holds that complex processes can best be dissected into a large number of very simple components, rather than a small number of relatively complex units (Watt 1966:3). The best level of complexity to be maintained in a model is a matter of some debate, some modelers preferring to lump information into as simplified a scheme as possible. To a certain degree these decisions depend on the quantity and quality of information available and on the modeling effort's objectives. A clear formulation of model objectives is essential to designing an appropriate model, as is the time horizon over which the model's behavior is expected to reasonably represent the real-world system.

Components of a system are often represented in flow diagrams by boxes, with causal relationships among them represented by arrows. Diagrams are an aid in organizing one's thinking about the system in preparation for translating these relationships into a *program*, or set of instructions to be followed by making the chain of calculations necessary to mimic changes in the system. The time-saving potential of a computer can be tapped at this point if the program is written as coded instructions in a language like FORTRAN.

Flow diagrams of system relationships, and programs built from them, are usually characterized by a number of *loops*, or closed pathways returning to their point of origin to influence their own behavior. Two types of *feedback loops* exist, with radically different results for system or model behavior. Positive feedback loops reinforce trends toward increase or decrease in the values of variables, while negative feedback loops tend to reverse deviations from a steady state. The result of a positive loop is for values of variables to either explode or collapse. A negative loop tends to damp oscillations in the value of variables, thus maintaining them at stable levels. Schematic representations of system relationships are themselves a form of model and can convey a significant amount of information about system behavior patterns. Difficulty in interpreting these diagrams increases as model structure becomes more complex. Various positive and negative loops may be acting on the same variable, the outcome of which can often be too complicated to foresee without actually performing the calculations as by running the corresponding computer program.

Once program structure is defined, quantitative data must be obtained to provide values for both the model's parameters and functional relationships. Functional relationships define how one variable varies in relation to variables or parameters causally linked to it. For the model to adequately represent the parts of a real-world system, both model structure and input values must be based on information from the field. Lack of adequate field observation is a major handicap for many simulation-based ecological studies. Parameters and functional relationships may be quantified in conjunction with development of model structure, as the investigator's perceptions of what factors are important change during the data collection process. Models are usually simplified and expanded to reflect better understanding of how the system functions derived from field observation and from information that becomes available on model behavior when simulations are performed using preliminary versions of the computer models.

Sensitivity testing is one way of using a simulation model to yield useful information both about model peculiarities and about the real system. A series of runs are made using alternative values for input parameters, and occasionally using alternative model structures. Behavior of state variables is observed under alternative assumptions to quantify the relative effect of, say, a doubling or halving of each parameter. From this information one can draw conclusions about the model's "robustness" to altered assumptions: a model may be capable of producing realistic results only within a narrow range of parameter values, since many of the negative feedback loops acting to stabilize the real-world system's behavior under extreme conditions may have been omitted to simplify the model. Sensitivity testing also reveals which parameters have the greatest overall effect on system behavior, improving the model's representation of the system by concentrating future research effort on improved measurements of these parameters and elaborating parts of the model where they play the greatest role. Identifying these parameters provides useful information about the real-world system, indicating, for example, system components that can most easily be manipulated for the benefit of humans. Factors can be identified with the greatest influence on increasing crop yields, reducing poverty, or reducing risk of colonist failures.

An additional vitally important step in assuring the relevance of simulated results to real-world systems is *validation* (Mankin et al. 1977), or comparing model behavior with what is known of the real system as a means of verifying the adequacy of the model as a whole. If clear differences are apparent in important aspects of model behavior within the range of parameter values in which the real system normally operates, one should doubt the reliability of results for parameter values outside these ranges, such as might be used in constructing hypothetical scenarios to test possible policy or management options.

The mathematical conclusions resulting from the model's behavior must be interpreted in terms of conclusions about the real-world system. This leap from the model to the real world requires an appreciation of the assumptions made in abstracting the model from the real system, weaknesses in input data and model structure, and differences between model and system behavior. Interpretation also requires awareness of the original purpose of the model.

Both in traditional research and research aided by simulation models, reaching a particular conclusion is not the end of the process. Rather, there is a feedback of the information gained from the study to the generation of new ideas, which will in turn be winnowed through the process of testing either by manipulating the system itself or by simulation. This continuous cyclical process forms the basis of scientific progress.

CHAPTER FOUR
Modeling the Agroecosystem of Transamazon Highway Colonists

Location and Description of the Study Area

A DETAILED LOOK at a limited area will illustrate many of the features of colonization in the Brazilian Amazon in general, although caution must always be applied in extending conclusions based on a limited geographical area. The area for the investigation into human carrying capacity, presented in more detail in chapter 5, consists of a part of INCRA's Altamira administrative area (PIC-Altamira) lying along the Altamira-Itaituba section of the highway. Within this an "intensive study area" of 236 lots has been delimited (figure 4.1).

The data collection effort was based in Agrovila Grande Esperança, located 50 km west of Altamira in the state of Pará. The agrovila is located in the município of Prainha at 3° 22' south latitude, 52° 37' west longitude (Brazil, IBGE 1978). The intensive study area includes the roadside lots from 15 km of highway between km 43 and 58 (by Highway Department (DNER) as opposed to INCRA kilometer measurement), and the full length of three lateral roads (15/17, 16/18, and 17/19) (figure 4.2).

The study area is situated in the interfluvial plateau of *terra firme* between the Xingú and Tapajós Rivers. According to the vegetation map produced by Projeto RADAMBRASIL, the intensive study area straddles four types of tropical rainforest.[1] Species lists of these vegetation types based on surveys by IBDF in the Altamira–Itaituba area (but outside of the intensive study area of the present study) indicate numbers of species ranging from 63 to 85 for trees with diameters at breast height of at least 25 cm. Wood volumes for the forest types range from 81 to 114 m^3/ha (Brazil, IBDF 1975:52–54). The IBDF data show considerable variation among one-hectare quadrats, confirming casual field observations. Variation in forest biomass not only affects the potential value of the wood, but also adds to the variability in labor requirements for clearing and possibly nutrient inputs from burning.

Topographic relief in the area is highly variable. A few colonists have nearly

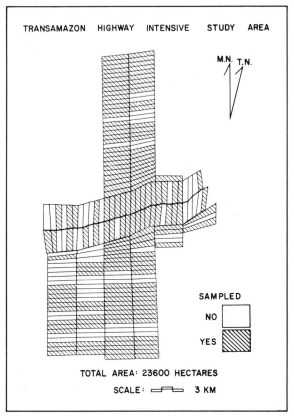

Figure 4.1. Map of the Transamazon Highway intensive study area showing sampled lots. Of the 236 lots in the area, 165 (70 percent) were sampled. The area is centered on Agrovila Grande Esperança, 50 km west of Altamira, Pará.

level lots, but most have severe limitations from steep slopes. Steep slopes not only preclude any prospect of mechanized agriculture, but also hold a considerable potential for erosion, especially under annual crops.

The soils of the area, like just about everything else, are very patchy. It is not uncommon for colonists to have several types of soil within their 100-hectare lots. Soils in the area include both some areas of *terra roxa* (ALFISOL), the best soil type, and larger areas of poorer soil types such as the ULTISOL "yellow latosol" (according to Brazil, RADAMBRASIL 1974, vol. 5).[2] The presence of large areas of the poorer soil types in the intensive study area, soils which are much more common on the Transamazon Highway as a whole than is *terra roxa*, make the soil and yield information from this

Figure 4.2. Poor transportation conditions in lateral roads have hindered the marketing of cash crops during the early years of colonization. (Lateral road 17/19, March 1975).

study more representative of the highway than are survey results in *terra roxa* areas. Most visitors to the Altamira area see only the agrovilas located on *terra roxa* at km 23 and km 90. Most researchers who have worked in the Altamira-Itaituba area have also focused on these relatively small areas of *terra roxa* (e.g., Homma 1976; Homma et al. 1978; Moran 1975, 1976, 1981; Smith 1976b, 1978, 1982). The main government agricultural research station in the area is also located on *terra roxa*.

The climate of the region is classified as Aw in the Köppen system (Pereira and Rodrigues 1971). A 36-year average of annual rainfall at Altamira is 1697 mm (Falesi 1972a:11). The intensity of the dry season varies greatly from one year to the next. The division between dry and wet seasons is nei-

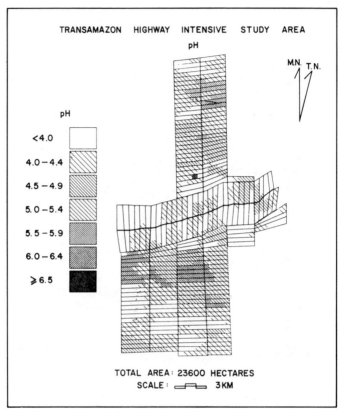

Figure 4.3. Map of soil pH under virgin forest in the Transamazon Highway intensive study area. Most soils are extremely acid, but variability is high.

SOURCE: Fearnside 1984e.

ther sharp nor predictable with precision. These facts add to the variability in burn qualities in agricultural fields. Weather stations maintained by EMBRAPA at km 23 and 101 west of Altamira show considerable differences between daily and even monthly rainfall totals.

Comparability with Other Tropical Areas

First, the representativeness of the study area as a sample of the Transamazon Highway as a whole must be considered. At this scale, as at any other, it is important to realize that there is no such thing as a "typical" area, since all areas have unique and important differences. The study area does, however, have many characteristics that are common to most of the area colonized on the Transamazon Highway. The soils are mostly of the poor ULTISOLs rather than the uncommonly fertile *terra roxa* (ALFISOL). There is no concentration of a single group of immigrants, such as *gaúchos* from Rio Grande do Sul or Japanese-Brazilian colonists. The area does not have government facilities or industries offering special opportunities for outside employment. The area is also far enough from urban centers to escape any special effect of marketing and employment opportunities that would not apply to most areas; although some produce is sold to supply Altamira (50 km away) and Agrópolis Brasil Novo (10 km away), most land use allocation is for crops to be sold to the government or to middlemen who appear at harvest time, rather than, for example, the weekly open market in Altamira. This is the most common situation on the Transamazon Highway, where distances between urban centers are great. The study area has also not had intensive promotion of single crops, such as the sugar cane growing area around 92 km west of Altamira.

As a representative of the Brazilian Amazon as a whole the study area has many commonly encountered characteristics, although the number of differences increases as one moves farther afield. The area has only *terra firme* (upland) rainforest, the most common class of natural vegetation in the Amazon, and has none of the *várzea* (floodplain) or *cerrado* (savanna) vegetation types. The poor soil of the study area is similar in fertility to that covering the vast majority of the Brazilian Amazon, but is not comparable, for example, with some areas colonized in Rondônia with markedly more fertile soil (although the amount of fertile soil in Rondônia is much less than is popularly believed). Difficulties with marketing, government bureaucracy, and transportation are surely typical of the region as a whole. The many types of assistance given by INCRA in the early years of the colonization program, which have since diminished, make the colonization experience here somewhat different from what can be expected in many present and future colonization initiatives. The very fact that the colonization in the area is directed colonization, rather than the more widespread pattern of undirected squatter settlement, makes the area atypical of pioneer areas in general. Violent strife over conflicting land claims has been relatively rare in the colonization area. Comparatively few lots have more than one family living on them, *moradores* ("dwellers") and *agregados* ("attached ones"), as is common in Rondônia.

As a representative of the humid tropics in general, the study area has some characteristics that are, and some that are not, generalizable to other areas. The open cash economy of the colonists is very common throughout the tropics, and is becoming even more so as isolated groups continue to be integrated into national and world economies. Many researchers have searched out systems as closed as possible, such as areas where shifting agriculture is still practiced in more or less traditional form. The present study is far more widely representative than these, although the added complexity of the system increases data needs and uncertainty. The uniform lot size of the colonization scheme is different from most settlement patterns elsewhere in the tropics, although social and economic class differences do exist within the colonist population and are continually increasing (Fleming-Moran and Moran 1978). Transportation, bureaucratic, and marketing problems are common to colonization areas throughout Latin America (Nelson 1973).

Agricultural problems such as poor soil, insects, weeds, and crop diseases are common to most of the humid tropics. The very acid soils of the study area are typical of the Amazon. These soils are different, however, from those in many tropical areas and present different obstacles to agricultural development. The study area soils are far poorer, for example, than many of the volcanic soils in Central America, or than the soils of the highlands of Papua New Guinea which are sometimes farmed under shifting cultivation with farmed times as long as 30 years. The study area's soils present different problems from the less-acid soils of much of tropical Africa. In Africa, burning can raise pH levels to the point where crops suffer from iron deficiency due to reduced availability under alkaline conditions (Sánchez and Buol 1975), whereas in the Transamazon Highway area burning inevitably results in better yields by minimizing such correlates of high acidity as aluminum toxicity and reduced availability of phosphorus. The great importance of burn quality discovered in the study area probably does not apply to areas where either soil acidity problems are less severe or where more favorable weather conditions result in better burns. At one location in Venezuela, for example, the slow decomposition of a very dense rainforest root mat is seen as the most important source of soil nutrients after clearing (Herrera et al. 1978), and by implication the ash from burning assumes a less critical role.

In comparing the Altamira area with other parts of the tropics it is also important to remember the effect of altitude in limiting agricultural, as well as natural, production (Janzen 1973b). Altamira's altitude near sea level (75 m in Altamira; approximately 100 m in the intensive study area) means that net production is higher than at high altitudes, but lower than the mid-altitude range where many human populations are concentrated in tropical areas. At low altitudes, high respiration both day and night results in lower net primary productivity (relative to the mid-altitudes), and lower surplus to support animals, including humans. Both human and crop diseases are greater problems at low than at middle altitudes. Other aspects of Altamira's physical environment, such as the extreme variability in rainfall (including both excessive rainfall and drought) are shared with some, but not all, tropical areas.

In summary, though no study area can be truly representative of either the Transamazon Highway, the Brazilian Amazon, or the tropics as a whole, the area chosen for the present study does have characteristics which are common to many other areas. The study area also has few characteristics which immediately identify it as atypical of the Brazilian Amazon region or of the development project of which it is a part. The Transamazon Highway

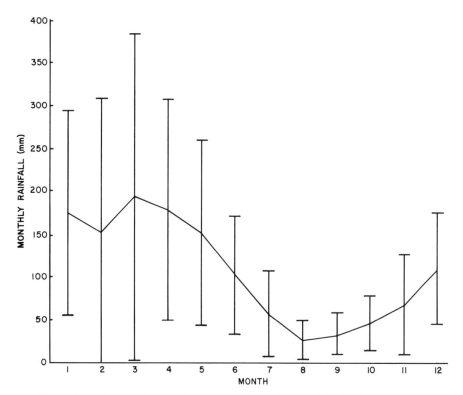

Figure 4.4. Mean and standard deviations of monthly rainfalls for Altamira (1931–1976).

SOURCE: Fearnside 1984f.

carrying capacity study was designed with the primary intent of answering fundamental questions about human carrying capacity, rather than as a survey to produce figures applicable to as large an area as possible. Although conclusions on carrying capacity cannot be simply extrapolated to other areas, many of the modeling features are transferable, even for areas with clear differences from the study area.

Methods

DATA COLLECTION

Field methods were refined throughout the project to arrive at a systematic procedure for data collection for the carrying capacity simulation programs. Informal interviews conducted with colonists and government officials in 1973 resulted in two questionnaires: one designed to extract relevant information on colonist origins, demography, consumption patterns, connections with the cash economy, and land-use decisions; the second to record the land-use and yield history of each field sampled. The questionnaires were revised during fieldwork to reduce the amount of information collected about product allocation and connections with the cash economy, and increase the detail of information on agricultural problems related to yields in the sampled fields. Conflicting information that could not be confirmed was discarded. Fieldwork lasted from May 1974 through August 1976, with twenty-six follow-up visits to the area between June 1978 and April 1985.

A cultural gulf separating colonists from government personnel must be bridged if reliable information is to be obtained. An important factor in bridging this gulf was frequent overnight stays in colonists' homes, necessitated by the long distance of some of the lots from the agrovila serving as the base of operations (some lay as far as 16 km away) since most of the study was conducted on foot. Colonists were given written summaries of the results of soil analyses done on their lots, which encouraged trust and cooperation.

Especially helpful was the colonists' good recall of details concerning agricultural problems and production, probably because almost all of the colonists were new to farming in the area. Farmers in areas of traditional agriculture, or in older pioneer areas, often cannot remember such information as how many sacks of rice were harvested from a particular field in a previous year. Because of bank financing and the use of annuals as cash crops,

Transamazon Highway colonists generally plant larger areas than do traditional subsistence farmers. The information on yields per hectare from a single harvest is much more reliable than when small amounts are harvested or when harvesting is done a little at a time.

During the last few months of the initial fieldwork, previously sampled locations were visited to assess soil changes. Information recorded with each sample included a location coding system and a grid for a rough map, to facilitate locating previous sample locations. Four field assistants did additional interviewing and information checking. A system of certainty codes allowed information gathered by assistants to be separated from that gathered by the author at any point during the data analysis (Fearnside n.d.[g]).

Informal observations and conversations with colonists were recorded in fieldnotes. This information provided estimates for a wide variety of items that could not be included in the formal questionnaires.

The intensive study area was carefully selected as a transverse slice of the colonization project in order to balance both market access difficulties, an important factor for lots far from the main highway, and soil quality (the better soils occur in bands roughly paralleling the main highway in this area). Great variability in the colonist population, the soils, topography, transportation conditions, and other features of the area makes bias from an unrepresentative sample an ever-present danger. At the same time, this consideration had to be weighed against the practical problems associated with pursuing a rigid random or stratified random sampling pattern. Difficulties of transportation and the importance of establishing a rapport with the colonists militated against such a procedure. Interviewers leaping unannounced from jeeps cannot expect to have the full cooperation of colonists whom they have never seen before; further, valuable sampling time would have been lost walking to random lots only to find that the colonists were not there. It was therefore decided to keep the numbers of lots on the roadside and in the various lateral roads as balanced as possible throughout data collection, but to capitalize on all opportunities to interview colonists and to sample lots. The high sampling density, with 70 percent of the lots in the intensive study area sampled, helped minimize the possibility of bias in the sample. Of the 236 lots in the intensive study area, 165 were sampled during the initial fieldwork used for estimating model parameters, with twelve more sampled outside of the intensive study area. In 131 of these, lot-level information was obtained from interviews with the colonist. The density of the sample also permitted modeling variation of soil qualities over small distances. The goal of the study—to elucidate underlying properties of tropical agroecosystems that affect carrying capacity—makes intensive sampling of a small geographical area appropriate.

Over 900 surface soil samples were taken to predict crop yields and changes in soils under different treatments. Samples were composites of cores taken with a tube-type soil sampler to a depth of 20 cm at a minimum of 15 locations in a field. Maps of initial (virgin forest) soil quality were made using both information from the 0-20 cm samples and data from equivalent depths from a series of 151 soil profiles,[3] each taken at one location using a soil auger, with samples analyzed at three to four depths in the 120 cm cores. Soil analyses were done in Belém, Pará, by the laboratory of CPATU (Center for Research in Agriculture and Cattle Ranching in the Humid Tropics). Laboratory methods, described in detail by Guimarães et al. (1970), were the North Carolina method for phosphorus; Kjeldhal method for nitrogen; Titurin for carbon; calcium and magnesium extracted with 1 N KCl and titrated with 0.025 N NaOH; potassium extracted with 0.050 N HCl and 0.025 N H_2SO_4 and determined in a flame photometer; pH determined in water with a potentiometer.

Some of the information on weather and financing was obtained from government sources. Information from these sources has not been used in other parts of the study. Data were obtained from fieldwork for soils, burns, erosion, and yields for rice, maize, *Phaseolus* beans, *Vigna* cowpeas, bitter manioc, and sweet manioc. Yield data for cacao, black pepper, pasture, and game, as well as much of the population sector data, relied heavily on the literature.

Data Analysis

A package of fifty FORTRAN programs and subroutines specifically designed for handling the data demands of KPROG2 was prepared (Fearnside n.d.[d]). In this package, routines related to crop yield prediction matched soil results with the appropriate data on yields, pests, treatments, weeds, diseases, density, interplanting, and so on. Calculations were made of days spent in various interplanting combinations directly from planting and harvest date information coded from the soil sample forms. Pest attack information was sorted, and attack intensity codes for different insect and vertebrate pests were tabulated by the stage in the crop's life cycle. Information related to weather, soil quality and changes, burns, erosion, and other factors was sorted and a variety of calculations made. Outputs are in a form that can be read directly by statistical packages.

Statistical analyses were performed using the Michigan Interactive Data Analysis System (MIDAS) (Fox and Guire 1976; Michigan 1976b). Data were stratified by certainty codes where appropriate, and cases with given pest attacks, varieties, diseases, interplanting combinations, and so on were excluded in selecting the final subsets of the data set for use in each analysis. Only a small fraction of the total number of cases and the variables pertaining to each case was used in any particular analysis. Most of the yield calculations relied heavily on regression, with regression on dummy variables used for burn quality effects. Burn qualities were predicted from weather data using discriminant analysis (Fearnside n.d.[c]).

Modeling Crop Yields

Annual Crops

Annual crops have been the agricultural mainstay in the first decade of Transamazon Highway settlement. Virtually all colonists plant annual crops, whereas only those few who have obtained financing plant perennials such as cacao and black pepper. Annual crops have not produced the high yields and agricultural surplus prophesied by government planners at the outset of the program. This so-called failure has been seized upon to justify dropping expansion plans for small farmer settlement projects in favor of aiding development of the region by large enterprises. In addition to agronomic problems, institutional obstacles in financing and marketing, and lack of transportation, have contributed heavily to the poor results (Bunker 1979, 1980a, b; Fearnside 1980b; Moran 1981; Wood and Schmink 1979). Agronomic problems, nevertheless, have also played an important role; the difficulties of the colonists are the sum of all contributing factors, both agronomic and social.

Upland rice *(Oryza sativa)* has been affected by many of the agronomic and other problems of agriculture on the Transamazon Highway. Despite poor returns, rice remains the most important crop for the majority of Transamazon Highway colonists. A closer look at how rice yields are modeled will demonstrate the kinds of agricultural problems encountered by colonists, and the relation of crop yields to other parts of the system. Later, the modeling of cacao, one of the perennial crops upon which planners and colonists are counting for sustained and profitable production, will point out additional problems likely to increase in importance as the area develops.

Rice yields are simulated in the carrying capacity models from a multiple regression, after appropriate assignments of values for independent variables. Subsequently, the yields predicted from the regression are adjusted for the effects of multipliers to represent losses from various agricultural problems excluded from the data set for the initial regression.

The first step in simulating yields is to assign initial values of 1.0 (indicating no loss) to the multipliers representing disease, toppling, rice variety, planting out of season, poor germination and interplanting with crops other than maize, manioc, or pasture (see table 4.1). Disease, toppling, and variety multipliers are assigned only once for any given lot and year. Rice planting density is assigned based on a mean density of 87.46 thousand hills/ha (with no pasture) (SD = 47.08, N = 255 fields). Multipliers for interplanted pasture, manioc, and other interplanted crops are assigned in accord with the land use of the patch. The multipliers for each agricultural problem have been calculated from data as the mean proportion of the regression-predicted yield obtained in fields affected by the problem.

If rice has been interplanted with maize (or with both maize and some other crop), the maize density is assigned based on a mean of 3,507 plants/ha (SD = 3444, N = 126 fields). Rice variety is assigned as hybrid (IAC-1246 or IAC-101), traditional *(canela de ferro)*, or other, based on frequencies of occurrence. Disease and toppling multipliers are then assigned using the frequencies of occurrence of these problems for each rice variety. Determinations of the occurrence of each of the agricultural problems listed in table 4.1 are based on the frequency of occurrence in the sample data.

The multipliers and the rice yield regression are derived from field data in an initial data set of 306 rice fields. Data falling into the following categories were eliminated as invalid: 1) data from fields with areas less than one hectare, which are less reliable than larger fields due to relatively larger errors in production and area estimates; 2) any data considered questionable due to contradictions or vagueness in colonist responses; 3) data based on yields estimated by colonists for rice harvested and piled but not yet threshed and sacked; 4) incomplete data for yield, area, density, maize density, or soil carbon, phosphorus, or aluminum; and 5) rice planting densities outside the range of values in experiment station studies, which were used in deriving expected "maximum" yields of each variety at different densities. Categories of valid data excluded from the regression were: 1) fields interplanted with manioc; 2) fields interplanted with pasture (no actual cases existed in the usable data categories); 3) fields interplanted with crops other than maize, manioc, or pasture; 4) fields with toppling reported; 5) fields with germina-

Table 4.1. Rice Yield Regression Excluded Category Summary

	Frequency		Rice Density (1000 hills/ha)			Effect on Yield (proportion of predicted yield)	
Condition	Percent	N (total)	Mean	SD	N	Mean	N (in category)
Manioc interplanted	10.1	306	93.05	44.59	28	1.080	19
Pasture interplanted	2.9	306	83.73	33.99	7	0.814	4
Other crop interplanted[a]	1.6	306	73.22	14.68	5	0.815	2
Poor germination	4.7	306				0.793	2
Disease							
Hybrid (IAC-1246, IAC-101)	1.9	211					
Traditional (*canela de ferro*)	12.0	25					
Other (nonbarbalha)[b]	14.3	5					
Lumped (nonbarbalha) effect[b]						0.806	5
Toppling							
Hybrid (IAC-1246, IAC-101)	13.3	211				0.760	7
Other (nonbarbalha)[b]	0.0	39					
Varieties planted							
Hybrid (IAC-1246, IAC-101)	84.4	183				1.000	c
Traditional (*canela de ferro*)	10.0	183				1.000	c
Other (nonbarbalha)[b]	5.6	183				0.823	
Planting Date Out of Season (November, March, April)	3.4	290				0.168	2

[a] Other than maize, manioc, or pasture.
[b] Barbalha variety rice, distributed by INCRA in 1973, is excluded. Barbalha resulted in widespread crop failure (mean yield = 21.4% of predicted IAC-1246 yield).
[c] By definition.

tion problems reported (no actual cases in usable data categories); 6) fields planted outside of the proper season (which includes December, January, and February); and 7) fields with rice varieties other than IAC-101, IAC-1246, or *canela de ferro*.

In cases where rice planting density is not known from direct field measurement, density is estimated using a regression on weight of seeds planted per hectare. When assessing soil effects, rice yields are expressed as proportions of expected yields interpolated from experiment station yields for the appropriate variety and planting density. In representing effects of soil characters, high nutrient levels are adjusted to represent the reduced effect of nutrients present in more than sufficient quantities. Soil carbon levels are adjusted to 2.0, meaning that values of percent dry weight carbon higher than 2.0 are assigned values of 2.0. This is in accord with the linear response and plateau model for predicting crop yields from soil nutrients (Waugh et al. 1975). The 2.0 percent value, used as the critical level for carbon throughout the study, is considered to be safely above the likely critical value (thereby underestimating the response to carbon), since a value of 1.2 percent carbon is considered high for crops in general in Brazil (Catani and Jacintho 1974:33–34).[4]

The phosphorus critical level used for purposes of simulation was 12.0 ppm (although none of the rice fields in the sample actually had soil phosphorus this high). This value was selected as safely above the value of 10.0 ppm considered high for crops in general (Brazil, IPEAN 1966; Catani and Jacintho 1974:33–34; Brazilian Soil Testing Service standards for Minas Gerais as cited by North Carolina 1974:149).

The regression for predicting rice yield (unhulled) is given by:

$$Y = 0.60\ A - 1.52 \times 10^{-5}\ B + 1.67 \times 10^{-2}\ C - 9.47 \times 10^{-2}\ D - 6.03 \times 10^{-3}$$

where:

Y = rice yield (proportion of experiment station yield)
A = carbon (% dry weight, adjusted to 2.0)
B = maize density (plants/ha)
C = total phosphorus (ppm, adjusted to 12.0)
D = aluminum (Al^{+++} in meq/100g)
($p < 0.05$, $r = 0.78$, SE $= 0.2029$, $N = 17$).

The relation given by this equation is applied in the simulation by first assigning experiment station yields in accord with previously assigned variety and planting density information. Predicted yield is calculated, and the standard error of the estimate (SE) is used to generate a predicted yield including effects of factors not specifically excluded from the data set for the regression. A maximum expected yield (in kg/ha) is calculated by multiplying experiment station yield (in kg/ha) by predicted yield (a proportion). Rice yield in kg/ha is calculated by multiplying the maximum expected yield (kg/ha) by multipliers for toppling, season, pasture, other interplanted crops, germination, and disease. Finally, an adjustment for technological improvement is made by multiplying rice yield by the value obtained for rice in separate calculations of technological improvements such as the increase in base yields through breeding programs.

Other crops have different agronomic problems. Maize, for example, is severely limited by rats (see table 4.2; Fearnside 1979d, and appendix) while disease is the greatest problem for *Phaseolus* beans (appendix), black pepper (Fearnside 1980a, and appendix), and cacao.

PERENNIAL CROPS

Cacao yields are predicted from a regression on soil pH based on data for yields in Trinidad (Chatt 1953, cited by Fonsêca et al. 1969) since cacao planted in the intensive study area was too young to be producing at mature levels at the time of fieldwork. Yield is predicted from the regression given by:

$$Y = 193.21 \, A - 744.29$$

where:

Y = cacao yield (kg dry seeds/ha/year)
A = soil pH
($p < 0.05$, $r = 0.73$,   SE $= 123.74$,
$N = 8$).

Although the range of pH values in the data set was 4.9–6.7, very high pH values in the simulation are adjusted to a critical value of 7.5, which is above minimal levels in several countries (Hardy 1961, cited by Fonsêca et al. 1969). Variability is introduced using the standard error of the estimate.

The cacao yields predicted using the regression based on Trinidad data are then multiplied by a correction factor to adjust for differences such as variety, climate, and shading practices that may result in different yield levels between two countries. The factor is derived by dividing the optimal (fertilized) yield of cacao expected on the Transamazon Highway of 1600 kg dry seeds/ha/year (Costa et al. 1973:25) by the maximum yield in the Trinidad data of 710 kg dry seeds/ha/year.

Table 4.2. Maize Yield Summary

Condition	Frequency (percent)		Maize Density (plants/ha)			Interplanted Crop Density (plants/ha)			Proportion Affected of Predicted Yield		
	Mean	N	Mean	SD	N	Mean	SD	N	Mean	SD	N
Maize alone	33.9	224	6274.8	4343.7	68	0		224			
Rice interplanted	51.8	224	3507.0	3444.1	126	126100	118280	41			
Manioc interplanted	13.4	224	5119.0	3530.3	21	5328.7	2647.0	18			
Other crops interplanted	9.4	224	5623.9	2794.3	18						
Poor germination	7.1	224							0.966		
Rats intensity 2 or 3	38.8	224							0.563	0.469	17
Disease	1.8	224							0.396		

Next, adjustment is made for age effects to represent the lowered yields in immature cacao. The proportions of the mature yield expected are: 1 year = 0.00; 2 years = 0.125; 3 years = 0.375; 4 years = 0.750; 5 or more years = 1.000 (Costa et al. 1973:25). Cacao in Amazonia has been estimated to have a productive life of 80–100 years (Morais 1974b:7.5); simulated cacao dies if it reaches age 90.

Disease effects are incorporated for blackpod disease *(Phytopthora palmivora)* and witches' broom *(Crinipellis perniciosa,* syn: *Marasmius perniciosus).* The probability of witches' broom (if established) attacking cacao is 0.13, and if not established is 0.03 (Fearnside 1978:335); the probability of blackpod (if established) attacking cacao is 0.29, and if not established is 0.11 (Fearnside 1978:339). Both diseases are now established in the study area. The probability of diseased cacao dying from witches' broom is 0.29, and of cacao dying from blackpod is 0.00 (Fearnside 1978:335, 339).

If a patch is attacked by one of these diseases, the yield will be reduced even if the trees in the patch are not completely killed. The multiplier for blackpod disease, if the patch is attacked, is 0.455, based on the mean proportion of healthy tree yields in variety trials in Ghana for five varieties tested (Dakwa 1974). For witches' broom, the value is 0.5, based on the assump-

Table 4.3. Data for Calculating Technological Change Factors

	Starting Annual Yields[a] (kg/ha)	Annual Increments in Yields (kg/ha)	Reference
Annual Crops			
Rice (with hulls)	1636	36.00	Páez and Dutra 1974:4.8–4.9
Maize (dry grain)	1418	18.00	Páez and Dutra 1974:4.8–4.9
Phaseolus (dry grain)	647	1.00	Páez and Dutra 1974:4.8–4.9
Vigna (dry grain)	647[b]	1.00[b]	Páez and Dutra 1974:4.8–4.9
Bitter manioc (flour)	3477[c]	32.89[c]	
Sweet manioc (flour)	3477[d]	32.89[d]	
Perennial Crops			
Cacao (dry seeds)	1600	24.00	Costa et al. 1973:25; Páez and Dutra 1974:4.8–4.9
Black pepper (dry seeds)	5500	0[e]	de Albuquerque et al. 1973:26
Pasture with animals (live weight gain of cattle)	10.6	1.74	Barcellos, 1974:6.18

[a] Fertilized or maximum expected yield for present varieties in the area.
[b] *Vigna* assumed same as *Phaseolus.*
[c] Converted to flour weight using 23% recovery (Smith 1976b:157).
[d] Sweet manioc assumed same as bitter.
[e] No increment value available for black pepper; zero assumed.

tion that the branches on a tree die at a constant rate once attacked while unaffected branches continue to produce at normal levels.

After calculating the cacao yield with all the above multiplier effects, a final adjustment is made for technological change, using technology multipliers derived for each cultigen by multiplying the number of years elapsed since the start of the simulation by the yield increment, adding this to the starting yield, and dividing the result by the starting yield (see table 4.3).

Land-Use Allocation Sector

Simulations of land-use allocation for subsistence and cash crops are simplifications of the real system, with only features judged most important being included. Allocation for experimentation with new crops, methods, or varieties is not included in the current version of the models, since the areas involved are small and yield improvements are modeled implicitly by technological correction factors in crop yield subroutines.

Subsistence allocations are based on areas of rice, maize, beans, and manioc needed to produce specified quantities for direct consumption by the simulated colonist's family, with a margin of safety for protection against poor harvests (appendix). A learning feature allows the simulated colonist to adjust amounts allocated, including the margin for poor years, based on the experience of colonists in the area (appendix). Consumption level standards, acceptable risks, and crop yields are specified or calculated in other parts of the program.

The cash crop allocation procedure used in the carrying capacity models, rather than following a "rational" decision procedure, is based on observed frequencies of different strategies among colonists of specific types. This empirical approach reflects observed characteristics of colonist behavior—the variety of land use mixes employed, for example, is far greater than one would expect were each colonist to select only the most profitable options indicated by a "rational" algorithm such as linear programming. Colonists on the Transamazon Highway are strongly influenced by profit expected from different allocation strategies, but the course finally chosen depends on such factors as the colonists' background, their own experiences and their neighbors' with crops in the area, opportunities for financing, and so on.

Cash crop allocations in the simulations are based on four possible strategies of lot development: annual cash crops, perennial cash crops, cattle ranching, and outside labor (see table 4.4). These land-use strategies are re-

Table 4.4. Land-Use Probabilities Based on Colonist Type

Colonist Type	Annual Cash Crops	Perennial Cash Crops	Ranching	Outside Labor	N
Entrepreneur	0.50	0	0.50	0	4
Independent farmer	0.55	0.30	0.15	0	27
Artisan farmer	0.79	0.17	0	0.04	24
Laborer farmer	0.85	0.07	0.03	0.05	67
N	93	17	8	4	122

Source: Fearnside 1980(b).
Note: Differences in land-use pattern: $p<0.001$, $\chi^2 = 28.6$, df = 9, N = 122

lated to four colonist types (Fearnside 1980b), patterned after the following typology devised by Moran (1976:38, 1979b): 1) *entrepreneurs*, who owned or managed land before arrival, had not migrated frequently, had urban experience, and owned many durable goods; 2) *independent farmers*, who were also owners and managers before arrival and had not frequently migrated, but had no urban experience and did not own many durable goods; 3) *artisan farmers*, who had not been owners and managers, had frequently migrated, had urban experience, and had owned many durable goods; and 4) *laborer farmers*, who also had not been owners and managers and had frequently migrated, but lacked urban experience and did not own many durable goods. In the present study colonists were assigned to these four types, but using only the criteria of previous experience as owners and managers and previous urban experience. No data were collected on durable goods or previous frequency of migration.

Table 4.4 shows the land-use patterns observed in the study area for the four colonist types. Simulated colonists choose different cash crop options based on probabilities of each choice among colonists adopting a given land-use pattern (table 4.5). Colonists of the entrepreneur type are the most likely to follow the ranching pattern, while laborer farmers choose annual cash cropping with the greatest frequency. Since 1976, the end year for these data, many more colonists have adopted the ranching strategy, and a somewhat smaller but still significant number of colonists have adopted the perennial cash crop strategy.

For both cash and subsistence crops the feasibility of allocating land to a particular crop is determined by labor and capital demands. Labor demands for agricultural operations, specified as to whether they must be performed by adult males only, must be calculated for each month of the year because of their highly seasonal nature. The value of family members for agricul-

Table 4.5. Probabilities of Cash Crop Land Use

											Sample Size	
	Rice	Maize	Phaseolus	Vigna	Bitter Manioc	Sweet Manioc	Cacao	Pepper	Pasture without Animals	Pasture with Animals	Cash Hectare Years	Colonist Years
Annual cash crops	0.66	0.09	0.07	0.00	0.07	0.01	0.00	0.00	0.10	0.00	411.3	85
Perennial cash crops	0.54	0.16	0.02	0.00	0.06	0.00	0.07	0.03	0.12	0.00	148.6	18
Ranching	0.35	0.03	0.05	0.00	0.05	0.00	0.00	0.00	0.40	0.12	187.4	10
Outside labor	0.83	0.00	0.17	0.00	0.00	0.00	0.00	0.00	0.00	0.00	9.0	4
										Total	756.3	117

SOURCE: Fearnside 1980(b).
NOTE: Cash crop areas are defined as areas over 0.5 ha for rice, maize, *Phaseolus*, and *Vigna*; over 0.2 ha for bitter manioc and sweet manioc; over 0.0 ha for cacao, pepper, and pasture.

tural work is adjusted for age and sex, and corrections are made for labor losses due to health problems, arranging for financing, and time spent in hunting and nonagricultural enterprises or outside employment (appendix). Hired labor can be substituted for family labor, capital permitting. Cash requirements are verified at the time of allocation, including seed purchase (if stored seed stocks are insufficient) and maintenance of crops already installed in the lot, as well as any fixed cash costs of the agricultural operations for the crop being considered. A labor and capital sufficiency check verifies that the total family labor, male labor, and capital supplies are adequate before the allocation is made. Though the current version of the models presupposes that colonists do not exceed resources in allocating land, such misjudgments do occur among colonists on the Transamazon Highway. Allocation decisions are made for each of the many "patches" into which the lot is divided for purposes of simulation, with the order determined in part by the ease of land preparation based on previous land-use category. The land-use allocation procedure used in KPROG2, including the values and justifications of parameters used, is presented in greater detail elsewhere (Fearnside 1980b).

Product Allocation Sector

Subsistence and Investment

Product allocation is done between consumption and investment, between investments in lot development and other ventures, and between durable and nondurable purchases. The first priority for simulated colonists in allocating production is feeding themselves and their families. Repayment of debts and investments come later. Subsistence allocations are made to consumption that fill the family's needs, or until product supplies are exhausted. A small allocation is made for seeds to be planted in the following year (appendix). Allowances are made for expected spoilage of a fraction of the stored product, different spoilage rates being used for products stored for later consumption or sale and the usually better protected products stored for use as seed (appendix). Marketing arrangements allow for the possibility that transportation to markets may be unavailable, depending on the location of the simulated colonist's lot with respect to the main highway and the time since the beginning of colonization (appendix).

Deficits in product needs not met by lot production are filled through purchases, to the extent that cash is available. A certain amount of cash is also needed for clothing and other nonfood subsistence items (appendix).

After cash has been allocated to subsistence, and any debts which the simulated colonist chooses to pay have been paid, a portion of the remaining funds is invested. The proportion invested depends on the colonist type—entrepreneurs and independent farmers invest greater proportions than do artisan farmers and laborer farmers (appendix). The remainder, including most of the simulated colonist's free capital, goes to consumption.

Colonist Diets

Colonist diets on the Transamazon Highway on the average are not bad by the standards of most developing areas. The observed average diet, with its nutritional value, is shown in table 4.6, together with the theoretical diet the colonist family would enjoy were it to consume the subsistence quantities of each product. That the observed values are lower than the theoretical values is due partly to the fact that there are more buyers than sellers of products in the population, and the theoretical subsistence diet is derived by taking the midpoint between consumption of buyers and sellers of each product (appendix). The difference is greatest for meat, the theoretical value corresponding to 50.2 kg/capita/year of meat, and the observed mean consumption to 17.7 kg/capita/year.

The theoretical diet that simulated colonists strive to attain is substantially above the standards used for carrying capacity criteria, which call for average per capita daily consumption of 2500 calories, 38 g of total protein, and 25 g of animal protein. This cushion protects many simulated colonists from failing to meet the standards, especially in the case of animal protein. The 50.2 kg/capita/year of meat in the theoretical diet is well above the approximately 45 kg constituting the minimum standard. The 25 g/person/day UN FAO standard used is itself quite high, 10 g/person/day or even less is considered adequate by many authorities (McArthur 1977). The higher figure is consistent with the standards used by local government organs (Brazil, ACAR-PARÁ 1974a).

Table 4.6. Colonist Diets

	Observed diet/capita			Nutrition/kg[a]			Nutrition/capita/day				Theoretical diet/capita[b]			
Item	Mean (kg/yr)	SD	N	Calories	Total protein (g)	Animal protein (g)	Calories	Total protein (g)	Animal protein (g)	Subsistence[c] (kg/yr)	Calories/day	Total protein (g/day)	Animal protein (g/day)
Rice[d]	116.6	64.4	19	2360	53.3	0	753.39	17.02	0	140.6	908.46	20.52	0
Beans[e]	43.2	19.0	13	3260	191.0	0	385.13	22.56	0	39.3	350.77	20.55	0
Manioc[f]	42.8	30.6	14	3440	13.6	0	402.91	1.59	0	61.7	574.35	2.30	0
Meat	17.7	10.9	11	1160	210.0	210.0	56.21	10.18	10.18	50.2	159.43	28.86	28.86
Tubers	135.4	54.1	4	1300	13.0	0	481.88	4.82	0	135.4	481.90	4.82	0
Purchases[g]							827.54	2.81	2.81		827.54	2.81	2.81
Totals							2907.06	58.98	12.99		3302.45	79.86	31.67

[a] Nutritional values from Brazil, ACAR-PARÁ (n.d.[a] [1974]). Corrected for waste.
[b] Target diet, not simulated results.
[c] Derived from midpoints between buyers and sellers of products.
[d] With husks.
[e] *Phaseolus* and *Vigna*, dried seeds.
[f] Sweet and bitter manioc flour.
[g] Purchase of products other than rice, maize, beans, manioc, game, and canned meat.

Population Sector

INITIAL POPULATION AND DEMOGRAPHIC PROCESSES

The size of the colonist population influences both demand for products and capability of lots to produce those products. When further increases in population size do not result in commensurate increases in production, consumption falls and failures occur.

Initial population characteristics generated in KPROG2 include the simulated colonist's type (see table 4.7), status as a game hunter or nonhunter, and the value of initial capital and capital goods (appendix). Age, sex composition, and health problems affect labor supply for agriculture (appendix).

The effects of nutrition on fertility have not been included in the simulation because in general they are indirect. Malnutrition is known to decrease fecundity (potential reproduction) only slightly, except in extreme cases of starvation (Bongaarts 1980). That fertility (actual reproduction) is positively correlated with malnutrition on a worldwide basis (Butz and Halbicht 1976) can be attributed in part to the more widespread use of birth control among the better nourished. In the simulation, fertility rates remain fixed at the 1970 rural Brazilian population levels used as parameters (appendix). This does not affect carrying capacity estimation done using fixed simulated population sizes.

Calorie and total protein consumption do affect mortality, however, when the simulated population size is not artificially frozen (appendix). The effects are derived from those used by the Club of Rome's Mesarovic-Pestle modeling team (Mesarovic and Pestle 1974a, b; Mesarovic et al. 1974) in the population sector of their multilevel world models (Weisman 1974; Ochmen and Paul 1974).

Other population processes modeled include colonist marriage, and individual and family immigration and emigration (appendix). Turnover in the population from replacement of entire family units is especially significant for the future of the area.

Table 4.7. Frequencies of Colonist Types in Original and Newcomer Populations

Settlement Status	Colonist Type				Sample Size
	Entrepreneur	Independent farmer	Artisan farmer	Laborer farmer	
Original	0.02	0.17	0.22	0.59	103
Newcomer	0.11	0.52	0.05	0.32	19

Colonist Turnover

The turnover of colonist families on the Transamazon Highway has been a major force in changing the landscape. Of 3,800 families settled in the Altamira Project, 665, or 17.5 percent, had left their lots by the end of 1974 (Brazil, INCRA 1974). If abandonments and sales were to continue at this rate, the original colonist population would have a half-life of only eleven years. Within the intensive study area, 33 percent of the lots were abandoned or sold at least once during the same 1971–1974 period. Since then most of the lots in the area have changed hands, with a major influx of prospective buyers in the area beginning in 1976. Land values have increased markedly, and regularization of land titles has made sales more profitable for many sellers. Since 1976, the number of lots abandoned by colonists has dropped to insignificant levels as lots can find ready buyers.

The reasons for sale and abandonment of lots are extremely diverse and have not been modeled explicitly in the carrying capacity simulations. On the Transamazon Highway, if lots do not produce enough to satisfy the colonists and better results are perceived as obtainable either on a new plot of virgin land or by moving to a city, the decision to abandon or sell a lot is a natural one. Since all colonists have given up a past life to come to the Transamazon Highway, the idea of being tied to a piece of land is foreign to them.

Many of the early abandonments can be ascribed to the shock of persons from other parts of Brazil and nonagricultural occupations trying to adapt to the hardships of pioneer farming in the Amazon. Abandonments for these reasons have already declined: those who "don't have the courage to face the conditions" (as the colonists put it) were weeded out, those that remain have become better adapted, and the conditions have improved. Although several abandonments can be attributed to undesirable conditions in the side road lots, just as many colonists left lots on the roadside in the intensive study area, possibly due to the greater ease of finding buyers for lots on the highway.

Abandonments appear to be random with respect to soil type within the intensive study area. However, one side road (No. 13/15), which has notoriously poor soil (outside of the intensive study area), was famed throughout the area for its large number of abandoned lots in the 1974–1976 period. Colonists with no previous agricultural experience appear to be more apt to leave than those who grew up as farmers: one former colonist who had sold popsicles (*picolés*) in the streets of Fortaleza in northeast Brazil before moving to the study area is now selling popsicles in the streets of Altamira.

Financial indebtedness is also a big contributor to colonist turnover. Many colonists borrowed large sums of money; a few squandered the funds, but more commonly the money was spent on agriculture as intended but led to a hopeless financial situation when crops failed. The poor burns in 1971–1972 and 1973–1974, plus the failure of the barbalha rice variety distributed by INCRA in 1972–1973, left many with large debts, the grace period for many of which expired on October 31, 1975.

Many colonists who would not abandon their land were willing to sell their lots, or to transfer them to anyone who would assume their debts. Before 1975 anyone buying a lot could assume the previous colonist's debts under the same favorable terms, but since then a prospective buyer must pay the entire debt at once. This change did not slow the turnover of colonists due to the increased inflow of prospective buyers, especially following the coffee-killing frosts in Paraná in 1975.

Actual bank foreclosures for failure to pay debts are very rare, although fear of foreclosure has often been a motive for sale. In spite of highly favorable financing terms, the existence of debts poses a constant threat to colonists. Even small debts with favorable repayment terms can be a problem if a crop fails and the colonist has nothing with which to pay. When a debt term extends over eight years or twenty years, it appears a virtual certainty that a crop will fail in at least one of those years.

Injuries and debilitating disease have been factors in several abandonments. One health-related factor sometimes mentioned by departing colonists is the bites of the *pium*, a small vicious, bloodsucking dipteran *(Simulium amazonicum)* appearing in clouds during the rainy months. Distribution of these black flies is very patchy, with some lots being much worse than others. Feet and ankles of a severely bitten person may swell up like balloons causing pain and other ill effects, making work impossible (Pinheiro et al. 1974a, b).

Many of the abandonments and sales stem from random events with no relation to lot production, such as marital or family problems, death or disablement of the head of the family, or quarrels with neighbors. Alcoholism has been a contributing factor in several cases, leading to fights, unwise use of money, lost work time, and two murders in the intensive study area.

Since virgin land is still available to all colonists in the intensive study area, none of the abandonments and sales so far are the result of declining production following environmental degradation of entire lots. Some of the abandonments and sales following one or more years of poor crops may have occurred under conditions similar to those expected with environmental degradation from exceeding carrying capacity. The large number of non-production-related abandonments and sales makes identification of cases linked to poor yields difficult.

Colonist turnover can be expected to play a continued role in changing agricultural patterns in the area. The influx of more capital with new settlers also represents a continuing supplement to the value of production generated within the area itself. Nevertheless, although the individual colonists may change, the processes of soil erosion, leaching, compaction, and regeneration, as well as the fixed sizes of the lots and of the colonization area, will remain the same.

The components of the Transamazon Highway colonists' agroecosystem, such as agricultural production, land-use allocation, product allocation, and population processes, must be woven together into a model of the entire system for the estimation of carrying capacity. The features of a stochastic model for simulating this system, and some of its results for estimating carrying capacity, are the subject of the next chapter.

CHAPTER FIVE
A Stochastic Model for Human Carrying Capacity on the Transamazon Highway

KPROG2 is a computer program that simulates the agroecosystem of a colonist population settled along a part of the Transamazon Highway. The model is designed to estimate sustainable carrying capacity under differing assumptions, carrying capacity being defined operationally as the population density at which the probability of colonist failure (i.e., falling below minimum consumption levels) exceeds a culturally defined acceptable maximum. The ultimate aim of such modeling efforts is to produce reliable carrying capacity estimates for use by development planners, thus forestalling the human suffering and environmental degradation that result from exceeding carrying capacity.

The present study examines carrying capacity only in the context of the types of agricultural systems currently in use or being contemplated for the Transamazon Highway colonization area. While allowances for technological changes are included, no consideration is given to nonagricultural technology that might, for example, support an urban center.

Features of the KPROG2 Carrying Capacity Model

KPROG2 is designed to reduce the bias in carrying capacity estimation as a consequence of restrictive assumptions such as those discussed in chapter 3. Flexibility is increased by supplying the maximum possible amount of information as input data rather than compiled programs. Features of the program intended to avoid the limitations of other carrying capacity estimation methods are discussed below. Further documentation can be found elsewhere (Fearnside 1979e; see also 1983b, n.d.[e]).

Multiple Limiting Factors

KPROG2 can make carrying capacity determinations based on multiple limiting factors. Colonist failure probabilities (proportions of colonist years in which minimal consumption standards are not met) are computed separately on the basis of calories, total protein, animal protein, cash per family, and proportion of land cleared. A combined probability of failure based on the per capita measures is also generated.

Dynamic Factors

Provision for technological change is made in two forms: gradual improvement of base yields of different crops, as from improved seed varieties, during specified year intervals; and changing land-use patterns—for example, a switch from annual crops to ranching or perennial crop strategies based on turnover in the colonist population.

A flexible population sector is included. A switch allows runs to be made with a "frozen" population at a fixed average family size so that different lot sizes can be simulated to compute sustainable probabilities of colonist failure at specific population densities. Carrying capacity is then estimated from plots of results from a number of runs, similar to the hypothetical curve presented in chapter 4. Alternatively, a dynamic population sector can be activated where demographic processes produce population changes.

Soil Quality Changes

Variable initial soil quality is generated from frequency distributions and Markov matrices of transition probabilities representing the probabilities of transition among fertility levels given moves between lots or between patches within a lot (Fearnside 1984e; see appendix, tables A.1 through A.6). The correlations among nutrients in actual virgin forest soils are maintained in the simulated soil qualities.

Burn qualities are variable, good and bad burns being predicted from cutting and burning dates (Fearnside n.d.[c]) and weather patterns generated to reproduce observed distributions for three types of variability: day to day variation, wet and dry years, and early and late rainy seasons (Fearnside 1984f; see appendix, tables A.7, 8, 9). Erosion is predicted for land in different uses from regressions based on slope and rainfall (Fearnside 1980d). Effect of erosion on soil quality is included, in contrast to many studies that ignore soil degradation.

Fallow times vary according to the pattern of actual colonists rather than being artificially restricted to periods corresponding to full recovery of soil quality. The circularity inherent in the many studies that assume a full fallow is thereby avoided. Runs with fixed fallow periods of different lengths can also be made.

Soil changes are computed and stored separately for each patch of land, creating a mosaic of patches in different stages of degradation and regeneration. Burn effects on soils are computed separately for three burn types: virgin, second growth, and weeds. Virgin and second growth burn effects depend on burn quality. Also included are days spent in different land uses and levels of other soil nutrients. Pasture soil changes are computed separately (Fearnside 1980c). Inputs from fertilizers and liming are included for cacao and black pepper with appropriate calculations of probabilities of fertilization, dosages, and cash adjustments.

AGRICULTURAL YIELDS

Crop yields are predicted with provisions for reproducing variability contributed by various causes. Crop yields are first predicted from regressions of yields on soil nutrients and other factors where sample sizes permit for upland rice, maize, beans *(Phaseolus vulgaris,)* cowpeas *(Vigna sinensis)*, bitter manioc, sweet manioc, cacao, black pepper, and cattle pasture (Fearnside 1979a, d, 1980a, 1984d). Planting density and interplanted crop densities are generated from observed frequencies and included as independent variables in regression models for rice, maize, and *Phaseolus* beans. Yields are first calculated as proportions of a base yield representing the yield for the crop from agricultural experiment station trials in the area. Factors not directly included in regression models, such as crop diseases, toppling, insect pests, vertebrate pests, and poor germination, are incorporated after the regressions through multipliers generated for these effects, expressed as proportions of the regression-predicted yield. The remaining unexplained variability is generated from the standard error of the estimate for the regression. Spoilage of stored products is also included.

Crop diseases are modeled for two cacao diseases, one black pepper disease, and one *Phaseolus* bean disease. Epidemiology of the diseases is reproduced to represent as realistically as possible the pattern in an area such as the Transamazon Highway. Crop diseases have repeatedly demonstrated the capacity to devastate large areas of these and other crops.

Animal protein sources are modeled with special care. Game obtained from hunting is harvested in accord with catch census by hunter or nonhunter status based on the actual frequencies of these two culturally distinctive types (using the data of Smith 1976b). Surplus game is sold to other lots within the simulated community. Animal protein is also obtained from chickens, which are fed on maize. Deficits not met from lot production and intra-community purchases of game are met by purchases of canned or dried meat or fish from outside, provided sufficient cash is available.

Land-use allocation includes upland rice, maize, *Phaseolus* beans, *Vigna* cowpeas, bitter manioc, sweet manioc, cacao, black pepper, pasture without animals, and pasture with animals. There are twenty possible crop combinations, plus four additional noncropped land uses.

Colonist Economic Behavior

Colonists are classed into the four types discussed in chapter 4. Initial values generated in accord with the colonist type include demographic information (appendix, table A.36), initial capital, and durable goods. Selection of four possible lot development strategies and four outside labor patterns is also based on probabilities specific to colonist type.

The variety of strategies gives great flexibility in representing the behavior of the colonist population. Newcomer colonists select among the possible strategy combinations with frequencies different from those of the original population. Product allocation between investment and consumption and between durable and nondurable purchases is also influenced by colonist type.

Labor supply is simulated so as to limit the amounts of land that can be cleared and cultivated in different crops (appendix, tables A.10–12). Supplements to family labor from hired hands are also included, with appropriate restrictions of the amounts of capital allotted by each family to investment in lot development. Labor supply is modeled to reflect the effects of several human diseases (appendix, tables A.39, A.41). The small but important probability of key family members falling ill at the time of felling, planting, or harvesting is also included.

Exchanges with the cash economy are modeled in detail. Selling and buying prices for products are variable (appendix, tables A.28, 29). Cash costs for installing and maintaining crops are included in the resource allocation sector (appendix, table A.13). Government supply of seeds is also included when appropriate. Interest, payment schedules, financing probabilities, and eligibility criteria are specified for twelve types of loans, with an inflation factor for repayment. While the Transamazon Highway area has received heavy government development subsidization, the bureaucracy and hidden costs of the credit system can represent a burden that often makes agriculture a money-losing venture for the individual colonists (see Bunker 1979, 1980a, b; Wood and Schmink 1979). Availability of transportation for taking cash crops to market is also included, since this was a major problem for many colonists on side roads during the early years of colonization. Provision is made for improvements of transportation conditions with time.

Nonagricultural income sources are included. Four types of outside labor, which often makes a critical difference to colonists, are modeled: daily wage labor, enterprises, government or professional employment, and working spouse or children. Simulated colonists can invest in small enterprises, such as general stores or pickup trucks.

Buffers against colonist failure are incorporated in the model at several points. Land-use allocation allows for the variability in yield for each crop with its implied probability of crop failure. The allowance is based on the "z" statistic of the colonists' maximum acceptable risk of failure (an input parameter) and the expected variability in yield for the crop. A learning function allows simulated colonists to base their decisions on cumulative past experience with crop yields in the simulated area, including both the mean and variance of those yields. Colonists' allocation for subsistence crops thus reflects past trends in crop yields (within limits of available labor, capital, and seed), providing a buffer against yield variability. When shortfalls occur for individual crops, the simulated colonist is able to buy subsistence items with money earned from cash crop sales or other sources, such as outside labor, sales of durable goods, or private loans if the colonist is lucky enough to get one. The diversity of crops planted also provides some measure of protection against a poor yield for any particular crop.

The subroutines of KPROG2's agriculture sector are incorporated into a smaller program, with myriad small but necessary modifications, to produce simulations of crop yields over time without the added complexity needed to translate these yields into human carrying capacity. The agricultural simulation, AGRISIM, requires that land-use decisions, and farmed and fallow times for annual crops, be entered from the terminal when each run is made.

Modeling Methods

The KPROG2 simulation is written entirely in G-level FORTRAN-IV using the Michigan Terminal System (MTS) as the operating system (Michigan 1976a). The simulations were run on the University of Michigan's Amdahl 470V/6 computer, which is software-compatible with the IBM 370.

A number of runs were made to test the sensitivity of simulation outcomes to changes in input parameters, although a full series of sensitivity tests has not been completed. The patch size (simulated small areas of land) used for storage of soil and land-use information, for example, was found to affect the land-use allocation sector, often making the difference between success and failure. After some experimentation, a patch size of 0.25 ha was selected as a reasonable compromise between economy of computer time and unbiased model behavior for all runs to be used in estimation of carrying capacity. For parameters such as patch size, which are purely arbitrary aspects of model construction rather than representations of real-world data, adjustments were made to avoid unrealistic behavior. Parameters representing the real world were introduced as is, without modification, in contrast to much of the existing modeling of human systems.

One of the switches in KPROG2 set interactively at the outset of each run selects between stochastic and deterministic runs. Deterministic runs provide a standard against which the stochastic results can be compared. The stochastic runs themselves are also completely reproducible by reusing the initial value entered as a "seed" for pseudo-random number generation.

Model Structure

Sectors and Causal Structure

The KPROG2 program may at first appear to be a vast labyrinth of sixty-three subprograms sharing information through sixty-two different labeled common regions. Actually, the essential causal structure of the program can be visualized quite simply from the arrangement of sectors into which its various parts are grouped. The agroecosystem acts as a filter, mapping information about initial resources into information about carrying capacity (figure 5.1). Inputs to the system in the form of variable weather, soil quality, colonist types, and so on enter into calculations that ultimately result in values for carrying capacity. The agroecosystem responds to initial conditions

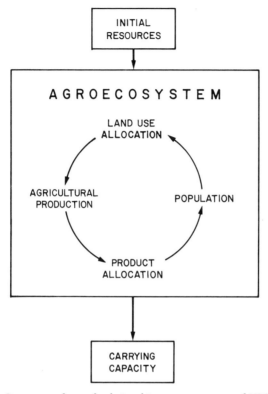

Figure 5.1. Summary of causal relationships among sectors of KPROG2.

within the context of the ecological and social processes included in the model, many of which buffer the colonist population from the effects of input variability.

The resource or land-use allocation sector models decisions related to how much land is planted in each crop and crop combination. These decisions naturally affect fallow periods and other items related to soil fertility. The agricultural production sector calculates how much of each crop is harvested based on areas planted, soil fertility, and a host of other factors influencing yields. The resulting production for the lot is then allocated among various possible uses in the product allocation sector, including consumption and investment in lot development. The amounts of products consumed contribute to maintaining the population when the population sector is in dynamic mode: population growth is sustained by adequate consumption, lesser levels of consumption leading to higher death rates. The population in turn influences land-use allocation; larger families have both increased capability for clearing land and a higher demand for subsistence crops. Information is taken from various points in this calculation process for computing carrying capacity, a calculation that is derived from, but not part of, the program.

In the agricultural production sector, weather affects soil quality through its influence on both burn qualities and erosion. Soil quality, in turn, is one of several factors affecting crop yields. Yields of the individual patches of land, when multiplied by the areas of the patches and summed for all patches in the lot, give the lot production information passed to the product allocation sector.

An indication of some of the major components and relationships included in the model is given by the causal loop diagrams (figures 5.2 and 5.3). The signs of the causal relationships are given at the head of each arrow, and in the case of the more simplified diagram (figure 5.2) the signs of some of the principal feedback loops are indicated in parentheses. Non-monotonic relationships, as well as categorical decisions influenced by stochastic processes, are also included.

In actually performing the calculations of the model, the logical sequence

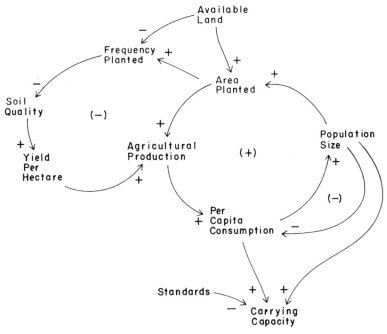

Figure 5.2. Simplified causal loop diagram of KPROG2.

SOURCE: Fearnside n.d.(e).

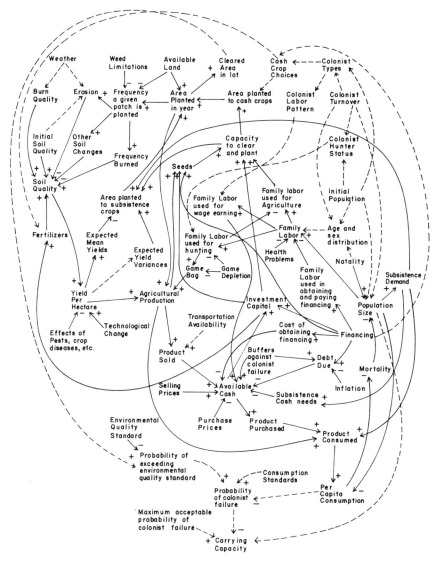

Figure 5.3. Causal loop diagram of relationships among some of the variables included in KPROG2.

Source: Fearnside n.d.(e).

depends partly on the geographic level to which the information applies. The order of events in the model is outlined briefly in the next section.

Sequence of Calculations

In program execution, subprograms are grouped by the size of unit to which they apply rather than by the sector in the program as defined by the principal causal relationships. The major iteration loops of the program (figure 5.4) group calculations into those done annually for each patch of land, each lot, and the entire community. Areawide statistics are computed for various measures after each year of simulation. Measures include both consumption information (such as calories, total protein, animal protein, and cash standard of living per capita), and indicators of environmental quality (such as proportion of land cleared and areawide averages for soil nutrient levels for land under different uses). Output of these measures permits calculation of carrying capacity based on the various criteria.

The operations in each major level shown in figure 5.4 are grouped by program sector in table 5.1.

Simulation Results

Standard Runs

Inputs for Standard Runs

A number of runs, made of both KPROG2 and the smaller AGRISIM program, were designed both to test effects of varying assumptions and to make carrying capacity estimates. The strategy for making a carrying capacity estimate is to run KPROG2 with a fixed population sector (for the parameter set used, the family size in each lot is always six persons). Runs are made with different lot sizes to achieve a range of population densities without distorting land use allocations, which depend on realistic family compositions for family labor calculations. (The alternative approach—varying family size with fixed lot sizes—will be discussed later in the chapter.) With the dynamic population sector activated, one can see how population trends are affected by both internal changes from births and deaths and from turnover in colonist population.

To simplify discussion, let us consider a single stochastic run with the

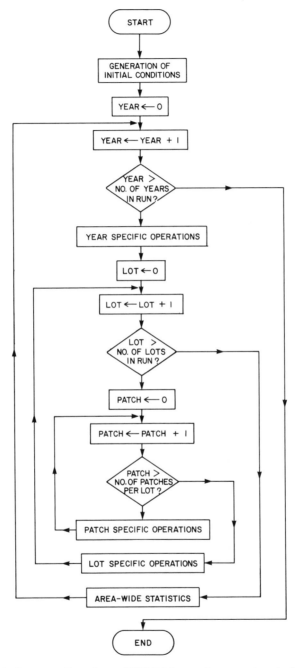

Figure 5.4. Summary flow chart of KPROG2 grouping operations by level.
SOURCE: Fearnside n.d.(e)

Table 5.1. KPROG2 Program Operations by Level and Sector

Level	Sector	Operation
Generation of initial conditions	Initial resources Population	Initial soil quality[a] Initial population (age-sex distribution) Initial colonist backgrounds Initial capital
Year-specific operations	Agricultural production Land-use allocation[a] Product allocation Population[a]	Weather generation Technological change Crop diseases[a] Strategy determination Seed needs determination Subsistence needs determination Hunting Wage labor and other income Financing Maintenance of perennial crops and pasture Land clearing Crop allocation Labor and capital sufficiency checks Prices Health Family labor equivalents calculation Newcomer population generation
Patch-specific operations	Agricultural production	Soils: burn qualities, burn effects, erosion, soil change, pasture soils, fertilizers Yields: rice, maize, *Phaseolus*, *Vigna*, bitter manioc, sweet manioc, pasture, cacao, black pepper
Lot-specific operations	Agricultural production Product allocation Population	Barnyard animals Transportation to markets Loan payments Buffers against failure Cash allocations Nutrition calculations Births and deaths Individual immigration and emigration Family immigration and emigration
Area-wide statistics	Carrying capacity	Probability of colonist failure Clearing statistics

SOURCE: Fearnside n.d.(e).
[a] The separate sets of loops that include these items do not appear in figure 5.4.

population sector frozen at six persons per family and a lot size of 25 ha, corresponding to a population density of 24 persons/km^2. Since the run described here is stochastic, the result represents only one of many possible outcomes for an area with this population density. Other outcomes can be generated by running the program with initial seed numbers for pseudo-random number generation different from the value used in this example (1113333).

The run was made using a community of ten lots with 100 patches per lot. Patches, the small hypothetical areas of land into which the simulated lots are divided, here correspond to 0.25 ha. No restriction was placed on the colonist types of the families occupying the ten simulated lots, nor on the fallow periods. Second growth in different age classes was cleared in accord with observed frequencies. In this run no change was assumed in base yields for crops through improvement in seed varieties, although the program has this capability.

The length of the run was 25 years. Because model behavior stabilizes well for small lot sizes within this period, longer runs would not produce substantially different results with the current data set. However, longer runs would discourage the short time horizon characterizing much development planning throughout the world, including Brazil.

Outputs of Standard Runs

Output from KPROG2 and associated graphics routines allows a ready visualization of land use, yields, and soil nutrient levels. The proportion of land allocated to rice, maize, rice interplanted with maize, *Phaseolus* beans, *Vigna* cowpeas, sweet manioc, bitter manioc, cacao, black pepper, and second growth are plotted. Areawide average yields are also plotted for each crop or combination. Average levels of soil pH, aluminum ions, phosphorus, nitrogen, and carbon are plotted for fields either bare (less than sixty days uncultivated) or in annual crops, and separately for fields in the other land uses. For these intermediate results, see Fearnside (1978).

The figures that follow present an example of the consumption measures used for calculating human carrying capacity.[1] Areawide averages for calories per capita (figure 5.5) indicate the average colonist's caloric intake is more than sufficient, not surprising given the ready availability of calorie sources from root crops. Despite high averages for the population as a whole, individual lots can easily fall below the 2,550 calories/person/day minimum specified in the input parameters (figure 5.6). Program output includes sim-

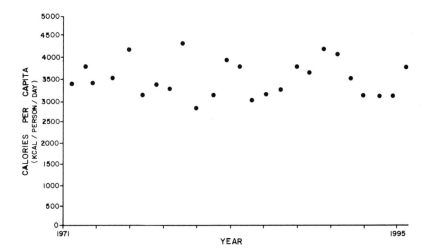

Figure 5.5. Simulated area-wide average calories per capita in example stochastic run of KPROG2 with population sector frozen at 24 persons/km².
SOURCE: Fearnside n.d.(e).

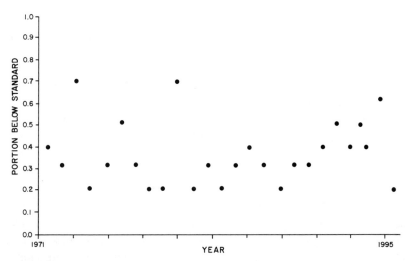

Figure 5.6. Proportion of lots below standard of 2,550 calories per capita per day in example stochastic run of KPROG2 with population sector frozen at 24 persons/km².
SOURCE: Fearnside n.d.(e).

ilar plots of areawide averages and proportions of colonists failing based on criteria for total protein, animal protein, and cash standard of living per family and per capita (Fearnside 1978). Because they consume large amounts of total protein, simulated colonists have low failure rates for this criterion. Simulated colonists consume more total protein than do actual colonists on the Transamazon Highway, even though the actual total protein consumption is high by the standards of many developing areas.

Simulated consumption of animal protein is more representative of the actual situation on the Transamazon Highway than is total protein consumption, but is likewise somewhat high due to the high priorities placed on animal protein in product allocation.[2] The unrealistically high simulated crop yields resulting from the crops' overdependence on pH in the model also contribute to the high simulated consumption levels.

Per capita cash standard of living provides an additional index. The minimum income used as a failure criterion was Cr75$54.40/person/month, or one-sixth of one minimum wage per month.[3]

Inclusion of environmental quality standards is regarded as a particularly useful feature of these models. Simulated colonists quickly clear all of their 25 ha lots in this run (figure 5.7). Within a few years clearing in all simulated lots exceeds the government-decreed maximum of 50 percent of the land area. Simulated colonists are not restrained from clearing beyond the legally permitted 50 percent since this law is not enforced by local authori-

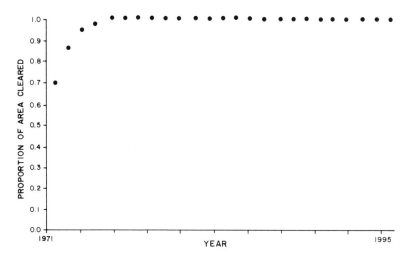

Figure 5.7. Proportion of total area cleared in example stochastic run of KPROG2 with population sector frozen at 24 persons/km^2.

ties on the Transamazon Highway—the 50 percent limit has no visible influence on the land clearing behavior of real colonists.

The importance of variability in production and consumption levels among lots is underscored by results of a number of stochastic runs (figure 5.8). Here proportions of colonist failures for individual years are plotted against areawide averages for calories for the same years. A significant proportion of colonists failed at areawide average values well above the minimum standards, reflecting an unevenness in product distribution among the colonists. Similar patterns were found for total protein, animal protein, cash per capita, and minimum wages per family (Fearnside 1978).

The failure probabilities from a number of runs made with the population sector frozen at different densities are needed for estimating carrying capacity. Sustainable probabilities of colonist failure have been calculated as the proportion of the total number of colonist-years in the last ten years of these 25-year-long simulations in which failures occurred by each criterion. Only the last 10 years of the run are used to calculate failure rates to allow behavior to stabilize before the rates are calculated. These failure probabilities are plotted against population density for five stochastic and eight determin-

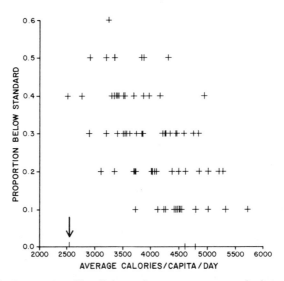

Figure 5.8. Proportion of lots below calories per capita standard versus area-wide calories per capita for years in several stochastic runs of KPROG2, showing the effect of variability among lots in consumption and production on probability of failure. Arrow indicates minimum standard used.

SOURCE: Fearnside n.d.(e).

istic runs for calories, total protein, animal protein, and cash per capita (figure 5.9). A number of runs would be needed at each population density to produce results that adequately reflect the effect of variation in crop yields and other factors. Such additional runs have not yet been executed.

The stochastic runs for calories and total protein resulted in higher colo-

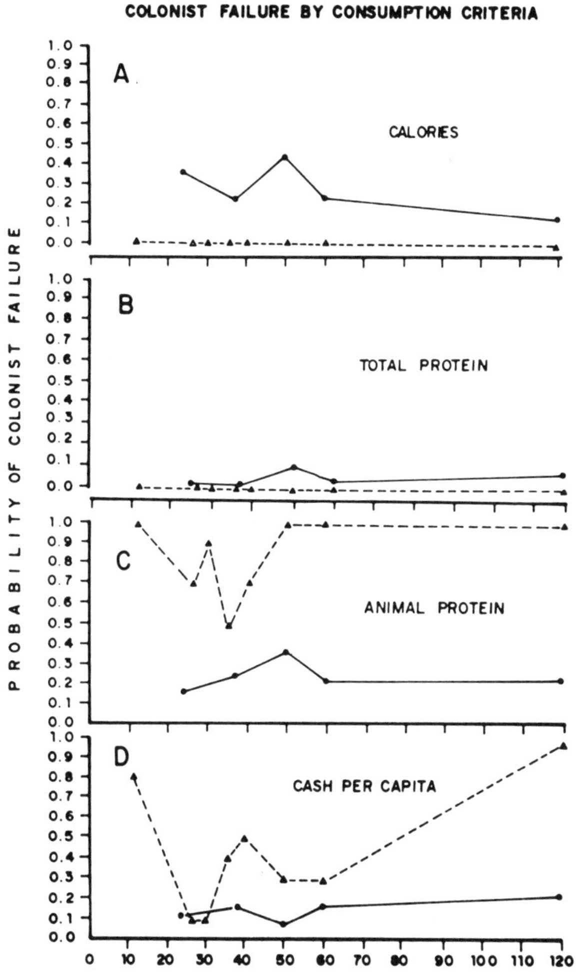

Figure 5.9. Colonist failure probabilities from 4 consumption criteria versus population density. Stochastic runs are indicated by circles and solid line; deterministic runs by triangles and dashed line. Failure criteria: a. calories, b. total protein, c. animal protein, d. cash per capita.

SOURCE: Fearnside n.d.(e).

nist failure probabilities than the zero probabilities for deterministic runs over all of the density range shown. Since these two criteria are the most easily satisfied through lot production, it is not surprising that corresponding failure probabilities are lower than those for animal protein and cash. The stochastic curves are higher, since some failures result when yields are allowed to vary more realistically. For animal protein and cash per capita, the failure rates for the deterministic runs are much higher than for calories or total protein. Production of animal protein in the lot from hunting or from chickens, which are fed on maize plus foraging in the colonist's yard, is usually inadequate, requiring supplements purchased with cash. When poor harvests result in restricted amounts of cash, the colonist cannot fill the need for animal protein through eating root crops or other readily available substitutes. At population densities above 40 persons/km^2 the annual failure rate in deterministic runs is equal to one. When variability in yields is introduced in the stochastic runs, good harvests are obtained by some colonists in at least a few of the years simulated, resulting in a lower rate of failure.

Deviations from Expected Results

The curves in figure 5.9 do not show the anticipated smooth increase in probability of colonist failure with increasing density. For animal protein and cash per capita in deterministic runs, the decline in the failure probabilities in the mid-range of population densities is attributable to the simulated colonists' ability to clear disproportionately large fractions of their lots in the first year. When these patches become uncultivatable due to competition from weeds the colonists will fail; later when the patches become available again for cutting and planting, the probability of exceeding the consumption standards increases. This pattern is attributable to the land use allocation procedure used in the simulation; actual colonists might not be so short sighted as to clear such a large fraction of their total available area in the first year. No evidence from the current study indicates that real colonists plan their fallowing schedules, although this possibility cannot be ruled out since the colonists in the intensive study area are not constrained by the small lot sizes of simulated colonists. At the lowest densities simulated, where colonists have enough virgin land available to reduce the impact of any synchronization in fallowing schedules, the high failure probabilities encountered are considered to be most realistic.

In stochastic runs some variation can also be seen between failure probabilities at different densities. Part of the variation undoubtedly is due to decisions being based on observed probability distributions rather than fixed pathways—were a different set of stochastic runs made at the same densities, different values for failure probabilities would be obtained. Increasing the number of simulated colonist-years would reduce the variability in the points shown in figure 5.9 which correspond to the proportion of failures in ten simulated colonists over ten years, or 100 colonist-years. Moderation of the fallow synchronicity effect may account for some of the lower failure probabilities at higher densities.

One reason for lower failure probabilities at the high density of 120 persons/km^2 is KPROG2's overdependence on pH as a predictor of crop yields, which leads to an exaggeration of crop-yield sustainability. In the very acid soils of the Transamazon Highway where poor burn quality, resulting in insufficient elevation of pH, has been a frequent problem in the first years of colonization, pH has indeed shown itself to be an excellent predictor of the yields of several crops. The effect of pH on crop yields is related both to correlations between pH values and several important nutrients and to the greater availability of nutrients at higher pH levels. Due to the small data sets available for predicting crop yields, the fact that, during the first years of colonization, soil pH overshadows the effects of other chemical, physical, and biological factors does not necessarily imply that other nutrients will not increase in relative importance in limiting crop yields over time, regardless of pH levels. That shorter intervals between cropping periods at high simulated population densities result in more frequent burnings for each patch of land, and consequently higher pH and crop yields, must be viewed as the result of deficient data for non-pH predictors. Altering the yield subroutines cannot be justified due to limitations of the current data set.

The range of densities shown in the simulated gradients also contributes to the deviation from expected trends. Ignoring for the moment the problems leading to unrealistically high yields, all of the curves could be expected to rise to a failure probability of one at some extremely high population. In the real world, the population density at which this probability is reached would probably be lower than these figures indicate. The question of greatest interest is: what becomes of the failure probability curves at the lowest population densities? For stochastic runs, the probability of failure appears to decline for cash per capita and animal protein at low densities, though the variability in results makes any firm conclusion on this point impossible without a large number of runs. One thing is clear: even with pH dependence and other features inherent in the program, failure probabilities even at the lowest simulated densities are quite high for most criteria. Even a failure probability of 0.1 or less per year implies a high probability of failing at least once over a span of several years. Also, when multiple criteria are used simultaneously, the probability that at least one of the standards will not be met is higher than the corresponding probability for an individual criterion. For example, in the stochastic run at 24 persons/km^2, the highest individual criterion-based failure probability is 0.36, while the combined probability of failure on the four per capita consumption criteria is 0.47.

Alternative Assumptions

Fallow Periods

Several assumptions were altered in different runs to judge impact on simulation results. Among these was the effect of fallow period, which can be controlled through alteration of the clearing probabilities for each age class of fallow land. A fallow period of six years was introduced into several deterministic runs at different densities by making clearing probabilities for all age classes less than six years equal to zero and clearing probabilities for all age classes six years or more equal to one. At high population densities a six-year fallow would result in failures during years when the large block of patches cleared in the first year was not eligible for planting. These high-density failure probabilities would not reflect real-world probabilities because of the effect of fallow period synchronicity on failure probabilities in the middle densities. At lower population densities this effect would pose less of a problem. At 12 persons/km^2, with the population sector frozen as before (corresponding to a lot size of 50 ha), the probability of colonist failure from animal protein is 0.2, as opposed to a probability of 1.0 for the free-fallow run. For cash per capita, failure probability is 0.2 for the six-year-fallow run and 0.8 for the free-fallow run. Calories and total protein both result in zero probabilities of failure in both the six-year fixed-fallow and free-fallow runs. Lower failure probabilities for animal protein and cash per capita in the six-year-fallow run may indicate the wisdom of longer fallows, which correspond more closely to the actual fallow periods in areas of traditional shifting cultivation. Reuse of young second growth or weeds as annual crop fields may be a short-lived phenomenon in the study area (Fearnside 1984g). As time passes, colonists may modify their agricultural behavior to match more closely the time-tested methods of traditional agriculturalists native to the Amazon region. Moran and Fleming-Moran (1974) have suggested such an acculturation process may be taking place among colonists coming from other regions of Brazil. As in the case of longer forced fallow periods, effects of such changes can be tested through simulation.

Colonist Types

The effect of altering the composition of the colonist population was examined. Changes in land use allocation brought about by turnover in colonist population have important implications for the future of the Transamazon Highway settlement and for conclusions regarding human carrying capacity (Fearnside 1980b, 1984g). The influence of colonist turnover can be seen from land-use information for four runs of KPROG2 (figure 5.10), showing differences among the colonist types distinguished in the land-use allocation sector (chapter 4).

For the two runs with laborer-farmer colonists only, the ten simulated lots in each run are initially occupied by laborer-farmers and all newcomer colonists are also laborer-farmers. In runs with all four colonist types, the original simulated colonists are replaced by newcomers before the inception of the final ten-year period represented in figure 5.10. For the first three years of this time period only 30 percent of the original colonists remain, resulting in a great increase in other colonist types at the expense of laborer-farmers.

In simulation runs at the 10-ha lot size (figure 5.10), the mixed population plants more of all of the groups of subsistence crops than do laborers: 15.2 percent more rice, 8.9 percent more maize (with a correction for planting density in the interplanted condition), 53.4 percent more "beans" *(Phaseolus* and V*igna)* and 13.4 percent more manioc (bitter and sweet). Except for rice, the same pattern holds with the 5-ha lot size—the mixed population plants 1.8 percent less rice, 22.6 percent more maize, 20.8 percent more "beans," and 28.6 percent more manioc than does the population of laborers.[4]

The small amounts of land allocated to cacao, pepper, and pasture by both laborer-farmers and the mixed population are unrealistically low, especially in the cases where the population is made up mostly of newcomers. The smaller size of simulated lots could be a partial explanation. More important, allocations during the early years of settlement on which the data for the program parameters are based differ from those made in subsequent years.

The colonist failure probabilities for runs with mixed types and with laborers only over the last ten years of the simulations suggest higher chances of failure for laborers by most criteria (figure 5.11). Values for combined probability of colonist failure, which represent the probability of any one of

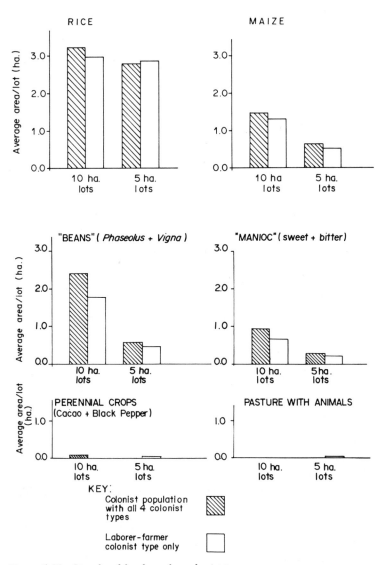

Figure 5.10. Simulated land use by colonist type.

SOURCE: Fearnside n.d.(e).

the four consumption criteria not being met, are more meaningful for carrying capacity than are those for individual criteria. The probabilities in figure 5.11 are quite high, undoubtedly due in part to the extremely high densities of the simulated populations in these runs. Although colonist fail-

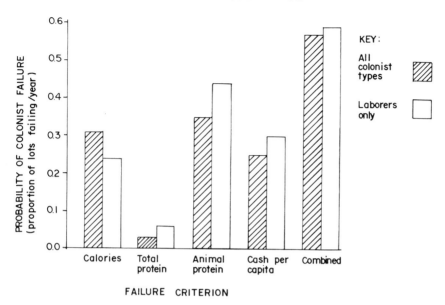

Figure 5.11. Colonist type and failure probabilities.

SOURCE: Fearnside n.d.(e).

ure probabilities at all densities simulated have been found to be higher than the standards of government planners, the failures are generally more frequent at higher densities, as would be expected from the relationship hypothesized.

Combined probabilities of colonist failure appear to increase with population density for stochastic runs with all colonist types and with laborers only in simulated lots of five and ten hectares (figure 5.12).[5] The differences in colonist failure probabilities between mixed populations and runs with laborers only in figure 5.11 may be in part related to the land use allocation differences discussed earlier, although the information is too scant for any firm conclusions. Differences in failure probabilities for colonists of different types could also be explained by the density differences between these simulation runs (figure 5.12). Although the results are insufficient to determine conclusively that failure probabilities are higher for laborers than other colonist types, the differences in land-use allocation in mixed population and laborers-only runs suggest that this may be the case.

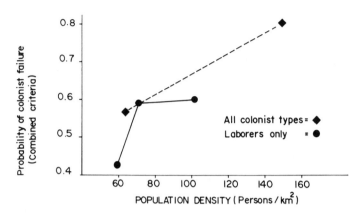

Figure 5.12. Combined probabilities of colonist failure at different population densities for colonist populations with all colonist types (triangles and dashed line) and with laborers only (circles and solid line).

SOURCE: Fearnside n.d.(e).

Conclusions on Carrying Capacity

The implications of these runs for carrying capacity depend on the critical value selected by the planner for the maximum acceptable probability of colonist failure. Such a value is implied by a land use classification system used by the Brazilian government's RADAMBRASIL Project in mapping the Amazon Basin. The RADAM report covering the Transamazon Highway study area classifies soil as "good" if the farmer could be expected not to fail more than once every five years (Brazil, RADAMBRASIL 1974 5:III/117). This corresponds to an annual failure probability of 0.13. Failure probabilities in stochastic runs almost always exceed this value for each individual criterion (see figure 5.9). Failure probabilities based on combinations of more than one criterion are even further above this value. Using a critical value of 0.13, the carrying capacity indicated would probably be below—possibly far below—the minimum population density simulated in the present stochastic runs of 24 persons/km². Further runs at these lower densities would be needed to confirm this conclusion. Since most of the deviations of simulation behavior from known conditions on the Transamazon Highway contribute to lower failure probabilities, refinement of the estimation techniques would probably result in lower values for carrying capacity.

It is clear that the high levels of variability observed in many parts of the agroecosystem are critical to failure probabilities, and hence to carrying capacity as operationally defined. The role of variability is shown by marked differences between the outcomes of stochastic and deterministic runs. The results of this study suggest that variability is a key factor that must not be ignored in seeking usable carrying capacity estimates.

The present study confirms the informal opinion of many that carrying capacity of tropical regions such as the intensive study area is very low for agriculturalists supported primarily on annual crops. This is not to suggest that development of these lands by larger enterprises, as in the conversion of extensive areas to cattle pastures, can provide the sustained yields necessary to support a human population at a higher carrying capacity. On the contrary, simulations of cattle pasture (Fearnside 1979a, 1980c) cast serious doubt on whether this form of production can produce such sustained yields.

CHAPTER SIX

Choosing Development Strategies for Rainforest Areas

Defining Development Objectives

CARRYING CAPACITY is inexorably linked to the formulation of development strategies. Government development objectives in the interest of both present and future generations can be divided into several categories, which will be discussed in the following sections.

SUSTAINABILITY

Long-term sustainability, both agronomic and social, of any system to be promoted is of paramount importance. Agronomic sustainability requires a reasonable balance of nutrients in the system, including compensation for losses from leaching, erosion, and nutrient export in the products harvested. Other requirements for continued productivity, such as control of soil compaction and energy from renewable sources, must also be met. Probabilities of destruction by pests or diseases must be low, and consideration should be given to an alternate land use should the present system fall victim to one of these biological problems.

Even if a crop can be sustained from the agronomic point of view, it may be short-lived for social reasons. Social sustainability requires that the system remain profitable over time. Fluctuations in yields and variations in market prices for inputs or the product can jeopardize long-term social sustainability. Also important is the practicality of enforcing regulations required for the system's functioning. The Brazilian tradition of selective application of laws to one's enemies, and of circumventing regulations through the ubiquitous *jeito* (Rosenn 1971), complicate planning greatly. In addition, social inequality and other underlying social conditions can affect a system's long-term sustainability. Sustainability is therefore linked with social forces arising from resource distribution and population pressure.

Unsubsidized Economic Competitiveness

Systems that depend on government subsidies for their survival often end up costing more than their intrinsic merits justify. Distortions introduced by such subsidies as tax incentives and low-interest loans have a way of becoming self-perpetuating even when the system has proven to be economically unfeasible. Cattle pastures in the Brazilian Amazon have followed this pattern (Fearnside 1979b). Systems must show themselves to be profitable in immediate (discounted) terms, thus obviating any need for governmental or other subsidies. This is not to imply that a complete laissez faire policy should be adopted; on the contrary, selective economic interventions can be used to promote social goals.

Maximum Self-Sufficiency

A balance must be struck between integration with the larger economy and local self-sufficiency. Dependence on imports of energy supplies, agricultural inputs, and basic food staples puts colonists at the mercy of price increases and the vagaries of supply. Small tropical farmers who produce a cash export crop with the intention of buying staples with the proceeds must contend with fluctuations in international commodity markets. Maximizing self-sufficiency is not to be confused with isolation from trade, which is necessary (within limits) for any agroecosystem. Some cash crops are needed to supply funds for purchase of goods not locally producible; the problem is to prevent loss of self-sufficiency in products that can be produced locally.

Self-sufficiency should include both economic self-sufficiency, meaning that enough is produced to satisfy demand, and nutritional self-sufficiency, meaning that enough is produced to satisfy nutritional needs of all regardless of purchasing power.

Fulfillment of Social Goals

To be viable, an agroecosystem must ensure minimum living standards, as measured by various criteria. Containing the probability of failure within acceptable limits for all criteria is one important condition. The amount of employment generated by different types of development must also be considered. On a regional scale, the cost of installing an agroecosystem can also be important if opting for an expensive development type means that social goals are not fulfilled elsewhere in the region.

Consistency with Development of Adjacent Areas

Development plans must ensure that adequate areas are available for ecological, Amerindian, and other types of reserves requiring intact forest. Boundaries of such reserves, once created, must be respected: development of surrounding areas must not create pressures to encroach on previously committed reserves. As with social sustainability, consistency with other uses is linked with resource distribution and population pressure.

Retention of Development Options

A prime consideration in selecting development strategies should be avoiding land uses that close the door to other possible uses. Unfortunately, the opportunity cost of destroying potentially renewable natural resources is rarely considered in making such decisions. The extensive destruction of the forest for cattle pasture is the most dramatic instance of short-sighted land use.

Minimal Effects on Other Resources

A minimization of adverse effects such as water pollution on other resources should be a part of any development planning. For example, fish breeding and feeding can be affected by the destruction of *várzea* (floodplain) and *igapó* (swamp forest) habitats. Some of the most important commercial fish in Amazonia depend on fruit production in seasonally inundated forests (Goulding 1980). The importance of fish as a protein source for the human population in the Amazon region means that any development with potentially deleterious effects on fisheries should be approached with extreme caution.

Minimal Macroecological Effects

Development strategies for rainforest areas must not be adopted without serious consideration being given to potential larger ecological effects. Species and genetic diversity, coevolved ecological relationships, and climatic stability are often sacrificed with little serious consideration given to applying effective, sometimes costly measures to curb these losses. Costs of ignoring these potential problems could be high, even though individuals and corporations enjoying the immediate profits of development will not pay the bulk of those costs, which may not come due for some time. The likelihood and magnitude of the potential effects are more than sufficient to warrant a careful examination of macroecological effects before a development strategy is instituted (see Fearnside 1985b).

Conflicts of Objectives

One frequent conflict of development objectives occurs between individual and societal interests. The classic formulation of this conflict is the "tragedy of the commons" analogy (Lloyd 1833; Hardin 1968): villagers would graze their animals in a common pasture and continue to add stock even when a further increase in the number of animals would lead to degradation of the pasture and lowered production; the fact that the full benefit of adding each additional animal to the pasture accrues to the individual, while the cost of lowered total production is shared by all the villagers, makes it advantageous to all villagers to increase their individual herds until the pasture is destroyed. The analogy applies directly to many common property resources such as fisheries, or the decisions regarding family size emphasized by Hardin (1968). The essence of the argument—the balance of individual gain with shared costs—also applies to many environmental problems such as the climatic impact of some development options in rainforest areas. Even if the total costs of these impacts should be far greater than the total gains from the developments, individual and corporate investors would continue to gain more than they lose.

The conflict between individual profit-seeking and the environmental and social concerns of society at large arises from a basic disparity between a system's sustainability and the investment patterns producing the highest economic returns. Investment decisions are made by comparing potential investments with returns obtainable from alternative investments in other parts of the larger economy (the latter summarized as a discount rate). Unfortunately, the rate of return that can be sustained by managing a renewable resource is limited by such biological factors as the growth rates of trees, which have no logical link with bankers' discount rates. If the discount rate is more than twice the rate of regeneration, as is often the case, it is to the investor's advantage to simply destroy the resource as quickly as possible and reinvest the profits in other enterprises (Clark 1973, 1976; Fife 1971).

The best-known example is the whaling industry: despite numerous studies showing that continued intensive whaling would lead to destruction of whale populations and an end to the industry, corporations continue to invest in whaling with the intention of simply scrapping equipment and reinvesting profits elsewhere later (Clark 1973). The problem is not a lack of knowledge, but the fundamental nature of economic decision making. This sad logic applies to many situations involving land-use decisions in rainforest areas—the question of whether a pasture or a silviculture operation is sustainable may be of minimal importance to the investor, even though lip service may be paid to sustainability as a development planning objective.

The question of discount rates poses a dilemma for development planning. High discount rates lead to overexploitation of potentially renewable resources, while artificially low rates lead to investment in economically inviable projects with poor returns. In the Brazilian Amazon, generous tax and fiscal incentives and subsidized interest rates have led to vast areas being felled for pasture despite the economic unfeasibility of this development strategy. The opportunity cost of capital is estimated at about 11 percent in most of Latin America by World Bank economists, and rates as high as 15 percent have been suggested as appropriate for Amazonian Brazil (Skillings and Tcheyan 1979:64). The actual discount rates used in economic calculations in the Amazon undoubtedly are far lower than the opportunity cost of capital, especially in the case of subsidized cattle ranching projects.

Skillings and Tcheyan have recommended using a system of shadow prices in calculating the costs of proposed projects (1979:65–66). Shadow prices for labor lower than market values and for foreign exchange and fossil fuels higher than market values would favor projects conforming to government goals.

Ideologies are irrelevant to the problem of sustainable economic uses of natural resources. Adam Smith's "invisible hand," the beneficent guiding force of traditional laissez-faire capitalism, has proved incapable of resolving the dilemmas posed by the tragedy of the commons and the discount rate problem. Likewise, the fundamental tenets of Marxism are inconsistent with renewable use of natural resources. Marx's labor theory of value, holding that the true value of any good is proportional to the amount of human labor required to produce it, leads logically to the conclusion that a natural resource like the Amazon rainforest has no value and therefore can be destroyed with impunity. The dilemmas posed by natural resource management require pragmatic solutions that cannot be derived from any ideology. A new approach to financial analyses is needed to make sustainable forms of development profitable and nonsustainable forms unprofitable.

Encouraging intensive development of small areas is one way of relieving pressure on the larger expanses of rainforest. The extremely low price of land in Amazonia is one impediment to development options requiring larger capital and labor investments to make land produce on a sustainable basis. Already deforested areas should be the sites for such intensive developments, not those still under native forest. The sustainable management of natural regeneration forestry requires a different set of economic guiding factors: low land costs and other capital demands. Amazonia needs to be developed in a mosaic of different patches of land, with different environmental quality restrictions and economic ground rules for each, along the lines of the ecosystem development strategy proposed by E. P. Odum (1969; Margalef 1968; see also Eden 1978; Fearnside 1979b, 1983c).

Colonization Programs and the Fulfillment of Objectives

SMALL-FARMER COLONIZATION VERSUS OTHER USES

Small-farmer colonization is only one of many development patterns competing for government funding and promotion. The changes in Brazilian government priorities in colonizing the Amazon, described in chapter 1, are testimony to the conflicting objectives of development policies. Most important are conflicts between social objectives and economic policies, e.g., producing agricultural products for export to other regions or countries (Fearnside 1984c). Most social motives for colonization are best satisfied through small-colonist settlements, in conjunction with other ancillary programs. Most of the other types of activities being promoted, such as cattle ranching and silvicultural plantations, require the capital and organization of large corporations. Appropriate objectives such as sustainability have not been prominent in planning decisions at any level. Whether small or large "colonists" are better in terms of these objectives depends entirely on the type of agricultural system promoted. Neither the annual crops favored by many small colonists nor the larger cattle ranching schemes have good prospects for sustainability (Hecht 1981; Fearnside 1979a, 1980c). Perennial crops, which have somewhat better prospects, can be grown in operations of a wide range of sizes. Other operations, such as sustained exploitation of native forest, do not pose insurmountable obstacles to the small colonist if cooperatives or other larger institutions provide the organizational structure and equipment.

The scale of development is related to differences in land-use allocation among colonist types within the settlement area on the Transamazon Highway. Smaller allocations of land to various crops by laborer-farmers as compared with other colonist types clearly leads to smaller agricultural surpluses and may also be linked to higher failure rates with respect to carrying capacity criteria. These differences among colonist types have important implications for policies governing selection of prospective settlers.

Objectives of colonization programs would have to be carefully assessed before recommendations could be formulated. If the principal objective of a settlement program is reducing the number of landless poor, then laborer-farmers would logically be given preference over other types regardless of consequences for failure probabilities and carrying capacity. If the measure of success for a program is production of an agricultural surplus for export, then the larger areas allocated to crops by nonlaborers would indicate these colonist types as preferable. The fundamental conflict between the objectives of agrarian reform and exportable production can be resolved only by an honest evaluation of each development program's objectives and probable consequences.

One possible objective for colonization programs, although not evident in the case of the Transamazon Highway, is the creation of self-sustaining communities capable of maintaining their populations at an acceptable standard of living without regard to benefits accruing to other parts of the country. If this should be the only criterion for evaluation, the independent-farmer colonist type would probably be the most viable choice.

Self-sufficiency rarely stands alone as a motive for colonization, often leading to the same fundamental conflict that applied in the case of the Transamazon highway. If a colonization project is to benefit not only individuals settled in the project but the Amazon region as a whole, the problem arises as to which colonist type is most appropriate. If an objective is to "effectively combat food shortages in tropical areas" (Moran 1976:98), then former owners and managers would be favored over laborers, as the land-use allocations simulated in chapter 5 indicate. Potential for conflict with other goals exists, however. Small-farmer colonization is currently being discussed as a remedy for immediate problems of: urban squatters in Amazonian cities, rural conflicts between squatters and large ranchers or speculators, and the recent flood of immigrants to the region from southern Brazil. The flood of immigrants is a combined result of the "pull" of opportunity in Amazonia and the "push" from concentration of landholdings for production of soya beans and sugar cane (encouraged for alcohol), the aftermath of killing frosts in coffee-growing areas, and continued population growth. These realities ensure continued pressure for the settling of laborers along with other colonist types.

Although the maintenance of population densities below carrying capacity should be a precondition for any colonization scheme, features designed to maximize carrying capacity, such as modification of selection procedures, must be weighed against all goals of the project under consideration. Land-use allocation patterns are at the heart of both the problem of carrying capacity estimation and attaining social objectives through colonization projects.

REDIRECTION OF DEVELOPMENT EFFORT

It has been proposed that development in rainforest areas in the Amazon be redirected to the central Brazilian scrubland, or *cerrado* (Goodland et al. 1978; Goodland and Irwin 1977; Goodland 1980a, b). *Cerrado* areas have lower opportunity cost if destroyed, are more resilient and better understood, and have better fertilizer response, less erosion, and fewer pest problems. *Cerrado* areas are also nearer to markets and have better transportation and other infrastructure available. Those resources devoted to Amazonian development with the intent of solving the problems of *other* regions might be better spent if applied directly to those regions. Not only is Amazonia incapable of solving the exponentially increasing problems of other regions, but applying resources directly to problem areas is more effective in the short run as well.

On a regional scale, it has been suggested that development efforts within the Amazon be concentrated in areas of second growth or degraded pasture (Rankin 1979), in the more fertile *várzea*, or that efforts be concentrated on increasing production in land already under cultivation rather than on expanding the areas in production. All these alternatives reduce motivation for felling rainforest, leave options open, and minimize both negative effects on other resources and adverse macroecological effects. The *várzea* is particularly attractive because of the annual renewal of its high soil fertility through flooding.

Efforts to produce higher and more sustainable yields on the *várzea*, second growth, degraded pasture, and already cultivated land would have to be combined with mechanisms to diminish the present strong motives for deforestation if pressure on rainforest areas is to be relieved. As of 1980 a new motivation has been added: a change in the structure of rural land tax laws in Brazil to tax "undeveloped" (uncleared) land at rates higher than those for "developed," land if more than a given percentage of a property is "undeveloped" and to increase the tax in successive years that land remains "unused" (Brazil, INCRA 1980).

Concentration of development in parts of Amazonia not presently under rainforest is consistent with a dual strategy for Amazonian development (Rankin 1979). Under such a strategy, short- and long-term plans would be separate, with present developments restricted to nonforested areas, thus giving time for researchers to develop sustainable and relatively nondestructive ways of making economic use of rainforest areas. The rainforest's potential could be expected to increase enormously by postponing the use of these areas. Two items are essential in any long-term plans for utilizing rainforest areas in Amazonia: 1) demarcation and defense of adequate permanent reserves of different ecosystems in the area, and 2) solution of the underlying problems leading to continued destruction of the rainforest. Carrying capacity estimation is a part of the solution.

Carrying Capacity and Development Policy

Human carrying capacity estimates should be central to the formulation of development policies of all types in tropical rainforest areas. Janzen (1972a, b) recounts some of the many examples of ecosystem destruction in the tropics linked to exceeding human carrying capacity, which can lead to lowered carrying capacity through destruction of "natural capital": "Such a gloomy scenario should not obscure the obvious fact that there are ways to determine human carrying capacity of a habitat without such countrywide or global destruction. These methods are where the first priority should be placed in funding" (Janzen 1972b:86).

Carrying capacity is a valuable concept around which to organize development. The very nature of carrying capacity requires long-range planning, a feature notably lacking from most of the development plans laid for rainforest areas. It also requires consideration of many diverse factors such as population density, levels of affluence, and income distribution, which are often treated as unrelated. The scales to which different objectives apply become explicit, whether rural farming populations, farmers plus urban centers within the region, or still larger units. In addition to its systems orientation, modeling carrying capacity focuses attention on the reality of limits, dispelling the illusion that infinite resources and agricultural potential exist. The potential for creating affluence and absorbing overflow from population growth and continued resource concentration in other regions is finite. Exponential growth within any area or region is incompatible with limits such as carrying capacity, even without the massive migration into regions such as the Brazilian Amazon.

Land tenure patterns are inseparable from carrying capacity. Failure rates for that portion of the population with fewer resources will probably be unacceptably high in terms of consumption-related measures of carrying capacity. With respect to environmental quality criteria, maintaining patches of land in uncut forest requires that land tenure status be defined and inequalities be reduced. In the Brazilian Amazon, land claims are established by deforesting and occupying a tract of land. The subsequent legitimization of these squatter claims after an area is settled provides a great motivation to deforest large areas, whether the "squatters" be poor *caboclos* or large enterprises. Laws aimed at maintaining the integrity of reserves and restricting deforestation are doomed to failure so long as this system prevails (Fearnside 1979b). The inequalities in land distribution that fostered this situation must be alleviated simultaneously with an end to the practice of legitimizing squatters' claims if proper forest management is to be instituted.

A concern for carrying capacity requires a development policy that considers all aspects of population: geographical distribution, age structure, rate of growth, and absolute size. Despite its size and resources, the Brazilian Amazon can no more sustain exponential population growth than can any other part of the world. Brown (1974:145) calculates that Brazil would have to double its 1974 food production by 1992 to keep pace with projected growth in demand, by its population, which was doubling every twenty-four years in 1970 and thirty years in 1980. Between 1948 and 1962, 84 percent of the increase in agricultural production in Brazil resulted from the cultivation of new land, only 16 percent from an increase in per hectare yields (U.S. DAgr. 1965:19, cited by Nelson 1973:21). Given the finiteness of the country, this pattern cannot continue forever. Areas of Amazonia such as Rondônia with its large influx of immigrants will be among the first to feel its effects.

Other countries have realized the need for a population policy as a part of development planning. In Egypt, a decision to slow population growth was made following the realization that the increase in food production made possible by irrigation from the Aswan Dam would be entirely absorbed by the increase in population in the Nile River valley during the period the dam was under construction (Brown 1974:143). In Mexico a parallel turnabout in national development and population policy took place in 1972 when it was realized that the country had become a net importer of food, the great strides in food production between 1955 and 1970 having been canceled by population growth (Brown 1974:174).

Brazil, however, is still experiencing a discrepancy between objectives and results. The most publicized objective for building the Transamazon Highway was that of alleviating overpopulation in northeastern Brazil, which had a population of approximately 25 million, growing at an annual rate of about 3 percent. This translates into a growth of 750,000 persons/year, or an average of about 2,055 persons/day. All three colonization projects on the Transamazon Highway settled a total of only about 5,000 families, which is equivalent to 30,000 persons at the average family size of six persons, or 14.6 days of population growth for the Northeast. Only 30 percent of colonists who settled in the Altamira area came from the Northeast (Brazil, INCRA 1974). Assuming that the other two areas had the same proportion of Northeasterners as Altamira, a total of 1,500 Northeastern families (about 9,000 people) were settled, or 4.4 days of population growth. Even if the percentage were higher for the other two areas, the fact remains that the colonization program did little to alleviate the population pressures in the Northeast. Population growth and land tenure concentration in rural areas of the Northeast continue more than a decade after the drought of 1970. Further, lack of planning to prevent populations of settlers from exceeding the carrying capacity of rainforest areas has led to environmental degradation and human suffering in development projects in many parts of the tropics (Dasmann 1972:788–89).

Planners and colonists alike look to the vast expanses of virgin rainforest in the Amazon for the panacea to any potential future problem—the "solution for 2001," as the Transamazon Highway was once dubbed (Tamer 1970). Exponential growth can quickly make a mockery of this logic, pointing up the urgency of determining the number of people who can be supported on a sustainable basis at a given standard of living and distribution of income without destructive environmental effects. Such knowledge is of much more fundamental importance than either projecting population and land-use changes or finding ways to increase agricultural production. One must assess what factors affect sustainable carrying capacities for humans and how to best obtain usable information about these factors. The time has come to take practical steps to avoid the human suffering that comes from exceeding carrying capacity.

APPENDIX
Summary of KPROG2 Model Parameters and Equations

Physical Conditions	161
Initial Soil Quality	161
Weather	163
Land-Use Allocation	165
Financing	165
Land Clearing	170
Subsistence Area Allowance	172
Colonist Learning and Allocation Behavior	172
Subsistence Margin Against Poor Yields	173
Modeling Crop Allocations	174
Interplanting Decisions	174
Labor and Capital Constraints	176
Cash Crop Choices	185
Modeling Soil Changes	186
Burn Qualities	186
Soil Changes from Burning	188
Virgin Forest Burns	188
Second Growth Burns	190
Weed Burns	191
Soil Changes Without Burning	192
Soil Changes Under Pasture	194
Fertilizers and Lime	195
Soil Erosion	199
Crop Yields	200
Maize	200
Phaseolus Beans	201
Vigna Cowpeas	203
Bitter Manioc	204
Sweet Manioc	204
Pasture and Cattle	205
Black Pepper	207

Product Allocation	209
Nonagricultural Income	209
Spoilage of Products	212
Transport to Markets	212
Buying and Selling of Products	213
Subsistence Needs for Agricultural Products	214
Subsistence Need for Cash	214
Debt Payments	216
Seed Requirements	217
Investment	217
Population Processes	218
Initial Population	218
Hunter Status	222
Family Labor and Health	222
Fertility and Mortality	224
Colonist Marriage	225
Immigration and Emigration	226

Model Parameters

KPROG2 model parameters and equations not included in the text are summarized in this appendix.

Physical Conditions

INITIAL SOIL QUALITY

Table A.1. pH in Initial Soil Quality Generation

Class	pH range	Frequency (%)	Mean pH	SD	N[a]
1	<4.0	33.0	3.7	0.1	389
2	4.0–4.4	30.2	4.1	0.1	356
3	4.5–4.9	15.3	4.7	0.1	180
4	5.0–5.4	12.5	5.2	0.1	148
5	5.5–5.9	5.3	5.6	0.1	63
6	6.0–6.4	3.6	6.3	0.1	43
7	≥6.5	0.1	7.1	(0)	1

[a] From a map of 1,180 quadrats based on 187 samples in virgin forest (Fearnside 1978:407, 1984f).

Aluminum content is calculated based on the pH and clay content values already assigned by:

$$Y = 11.43 - 7.68 \ln A - 6.27 \times 10^{-2} B \qquad \text{Equation A.1}$$

where:

Y = aluminum (Al^{+++} in meq/100g)
A = pH
B = total clay (%)
($p < 0.001$, $r = 0.73$, $SE = 1.56$, $N = 118$ virgin forest samples).

Nitrogen is then calculated from the carbon and pH values for the patch from:

$$Y = 0.132 A + 2.20 \times 10^{-2} B - 0.120 \qquad \text{Equation A.2}$$

Table A.2. Transition Probabilities for Virgin Soil pH

Beginning pH Class	Ending pH Class							Sample Size
	≤3.9	4.0–4.4	4.5–4.9	5.0–5.4	5.5–5.9	6.0–6.4	≥6.5	
	Move of 100 meters[a]							
≤3.9	0.00	0.33	0.00	0.33	0.33	0.00	0.00	3
4.0–4.4	0.09	0.73	0.09	0.00	0.00	0.00	0.09	11
4.5–4.9	0.00	0.17	0.33	0.00	0.50	0.00	0.00	6
5.0–5.4	0.09	0.00	0.00	0.36	0.36	0.18	0.00	11
5.5–5.9	0.11	0.00	0.33	0.44	0.00	0.11	0.00	9
6.0–6.4	0.00	0.00	0.00	0.40	0.20	0.40	0.00	5
≥6.5	0.00	1.00	0.00	0.00	0.00	0.00	0.00	1
	Move of 500 meters[b]							
≤3.9	0.40	0.43	0.09	0.06	0.00	0.00	0.03	35
4.0–4.4	0.25	0.43	0.15	0.08	0.08	0.02	0.00	61
4.5–4.9	0.12	0.36	0.24	0.08	0.12	0.08	0.00	25
5.0–5.4	0.12	0.29	0.12	0.24	0.24	0.00	0.00	17
5.5–5.9	0.00	0.25	0.15	0.20	0.30	0.10	0.00	20
6.0–6.4	0.00	0.20	0.40	0.00	0.40	0.00	0.00	5
≥6.5	1.00	0.00	0.00	0.00	0.00	0.00	0.00	1

Source: Fearnside 1984e.
[a] Calculated from samples which are 100 meters from each reference sample ± 100 meters. Used in simulation for moves between patches within a lot.
[b] Calculated from samples which are 500 meters from each reference sample ± 500 meters. Used in simulation for moves between lots.

Table A.3. Clay for Initial Soil Quality Generation

Class	Range (% dry weight)	Frequency (%)	Mean Clay (% dry weight)	SD	N[a]
1	0–14	21.6	9.71	2.83	225
2	15–29	25.3	21.70	4.05	299
3	30–44	23.6	37.95	4.48	278
4	45–59	21.2	50.00	4.00	250
5	60–74	7.5	69.72	3.57	88
6	75–89	0.8	78.80	1.03	10

[a] From a map of 1,180 quadrats (Fearnside 1978:427; 1984e) based on 200 samples.

where:

$$Y = \text{nitrogen (\% dry weight)}$$
$$A = \text{carbon (\% dry weight)}$$
$$B = \text{pH}$$

($p < 0.001$, $r = 0.86$, SE $= 0.030$, N $= 53$ virgin forest samples).

Table A.4. Slope for Initial Soil Quality Generation

Class	Range (slope in %)	Frequency (%)	Mean Slope (%)	SD	N[a]
1	0–4	32.0	2.0	1.4	378
2	5–9	18.7	6.3	13.8	221
3	10–19	29.3	13.8	2.5	346
4	20–29	13.5	22.8	2.2	159
5	30–39	5.8	37.8	2.9	68
6	40–69	0.6	57.9	3.3	7
7	70–90	0.1	89.0	(0)	1

[a] From a map of 1,180 quadrats (Fearnside 1978:437; 1984e) based on 225 measurements.

Table A.5. Carbon for Initial Soil Quality Generation

Class	Range (% dry weight)	Frequency (%)	Mean Carbon (% dry weight)	SD	N[a]
1	<0.50	1.9	0.39	0.08	22
2	0.50–0.86	31.0	0.70	0.11	366
3	0.87–0.99	5.1	0.94	0.03	107
4	1.00–1.49	40.8	1.18	0.13	482
5	1.50–1.99	17.2	1.70	0.14	203

SOURCE: Fearnside 1984e.
[a] From a map of 1,180 quadrats (Fearnside 1978:423; 1984e) based on 75 samples.

Table A.6. Phosphorus for Initial Soil Quality Generation

Class	Range (ppm)	Frequency (%)	Mean Phosphorus (ppm)	SD	N[a]
1	0–1	83.8	1.0	0	989
2	2	8.3	2.0	0	98
3	3–4	5.7	3.0	0.1	67
4	5–6	2.1	5.2	0.4	25
5	7–9	0.1	7.0	0	1

SOURCE: Fearnside 1984e.
[a] From a map of 1,180 quadrats (Fearnside 1978:409; 1984e) based on 187 samples.

WEATHER

Rainfall, evaporation, and insolation must be simulated for use in calculating erosion and burn quality. Monthly rainfall is generated for the planting season (January through May) by generating a season total from the mean seasonal rainfall of 1396.2 mm and standard deviation of 300.9 (N = 21), and then multiplying this total by the proportion of the seasonal total expected for each month.

Rainfall values are assigned separately for the months of June, July, and August, using the mean and standard deviations for rainfall in these months (table A.7). The burning period (September through December) is treated in the same way as the planting period, using the proportions of seasonal totals (table A.7) and a value for total rainfall in the period generated from the mean of 281.7 mm and standard deviation of 148.1 mm (N = 12).

Monthly evaporation figures are calculated from:

$$Y = 102.0 - 0.150 \, A \qquad \text{Equation A.3}$$

where:

Y = monthly evaporation (mm)
A = monthly rainfall (mm)
($p < 0.0001$, $r = 0.57$, SE $= 32.06$, N $= 45$ months).

Variability is introduced using the standard error of the estimate (SE) for the regression.

Monthly insolation is then calculated using a similar regression on monthly rainfall.

Table A.7. Monthly Rainfalls as Proportions of Period Totals

Month	Mean	SD	N
Proportions of rainy period (Jan.–May) totals			
Jan.	0.1778	0.0588	21
Feb.	0.2141	0.0880	21
March	0.2631	0.0502	21
April	0.2136	0.0505	21
May	0.1312	0.0522	21
Rainfalls in mm for months not included in periods			
June	77.58	48.18	30
July	58.82	50.74	28
Aug.	28.07	22.39	27
Proportions of burning period (Sept.–Dec.) totals			
Sept.	0.1862	0.1099	12
Oct.	0.1995	0.1075	12
Nov.	0.1850	0.1377	12
Dec.	0.4293	0.1026	12

SOURCE: Fearnside 1984f.

$$Y = 156.68 - 0.180 \text{ A} \qquad \text{Equation A.4}$$

where:

Y = monthly insolation (hours)
A = monthly rainfall (mm)
($p = 0.0002$, $r = 0.52$, SE $= 44.05$, N $= 45$ months).

Land-Use Allocation

FINANCING

Twelve types of financing are included in the program: 1) land purchase and house loans; 2) virgin forest felling (including the underclearing *(broca)* and piling of unburned material *(coivara)* with the eight-year loan term of the original plan (which was in effect from 1971 through 1974); 3) virgin forest felling loans with higher interest and one-year terms which came into effect in the 1974–1975 agricultural year; 4) INCRA debt for salaries and nondurable items bought on credit during the first months of the colonization program; 5) other debts for durable items (such as power saws); 6) INCRA seeds; 7) rice *custeio* (planting, weeding, and harvesting); 8) maize *custeio*; 9) *Phaseolus* bean *custeio*; 19) perennial crops (pepper or cacao); 11) cattle and pasture with animals (including fencing); and 12) private loans, including credit from colonist-owned shops in the agrovilas. The frequencies, amounts, and terms of the 12 loan types are shown in table A.9.

Table A.8. Variability in Daily Weather as Proportion of Monthly Totals

Month	Precipitation		Evaporation		Insolation	
	SD	N	SD	N	SD	N
Jan.	0.0485	142	0.0219	123	0.0243	122
Feb.	0.0490	113	0.0178	113	0.0307	85
March	0.0412	159	0.0123	153	0.0261	123
April	0.0479	150	0.0270	150	0.0289	120
May	0.0696	155	0.0258	155	0.0667	124
June	0.0648	150	0.0302	150	0.0312	120
July	0.0843	155	0.0302	154	0.0162	124
Aug.	0.1096	147	0.0267	146	0.0108	124
Sept.	0.0969	120	0.0289	120	0.0136	120
Oct.	0.1013	119	0.0250	119	0.0231	69
Nov.	0.1346	120	0.0150	120	0.0263	120
Dec.	0.0677	124	0.0302	124	0.0336	124

NOTE: Means used are monthly totals divided by number of days in month.

Table A.9. Financing: Frequencies, Amounts, and Terms

Loan Type	Probability of Financing[a]	N	Hectares Financed Mean	SD	N	Amount per Ha Mean[b]	N	Beginning Year	Ending Year	Grace Period (yrs.)	Loan Term (yrs.)	Interest (%/yr.) without Penalty[c]	Interest (%/yr.) with Penalty[d]
1. Land and house	1.0[e] or 0.0[f]	N[g]	1[h]	0	N[g]	46,273	N[i]	1	999[j]	3	20	6.0	6.0
2. Virgin clearing type 1	0.79	67	6	0	N[g]	1,037	N[k]	1	3	3	8	7.0	12.0
3. Virgin clearing type 2	0.74	21	6	0	N[g]	450	N[l]	4	999[j]	0	1	10.0	13.0
4. INCRA debt	1.00[e] or 0.00[f]	N[g]	1[h]	0	N[g]	7,744	N[m]	1	999[j]	1	4[n]	6.0	6.0
5. Durables	0.12	N[o]	1[h]	0		7,588[p]	N[q]	1	999[j]	3	8	7.0	12.0
6. Seeds	1.00	N[r]	3	0	N[r]	202	N[s]	1	4	0	1	6.0	6.0
7. Rice	0.73	78	3	0	N[r]	430	N[t]	1	999[j]	0	1	10.0	13.0
8. Maize	0.67	24	3	0	N[r]	190	N[t]	1	3	0	1	10.0	13.0
9. *Phaseolus*	0.57	28	2	0	N[r]	340	N[t]	1	3	0	1	10.0	13.0
10. Perennial crops	0.50	10	2	0	N[u]	8,748	N[v]	1	999[j]	3	8	10.0	13.0
11. Pasture	0.05	21	5	0	12	1,156	12	1	999[j]	3	8	10.0	13.0
12. Private	0.10	N[w]	1[h]	0		1,000	N[w]	1	999[j]	0	1	0.0	0.0

SOURCE: Fearnside 1980b.

[a] Probabilities of financing being requested *and* picked up at the bank given a colonist has decided to use the crop in question as a cash crop and is solvent.
[b] Amount financed in Cruzeiros of 1, January 1975. For loan types 1, 4, 5, and 12 this is the amount of the loan. (All amounts corrected for inflation at 35%/year. Minimum wage = Cr$326.40/month; US$1 = Cr$7.4).
[c] Without penalty for late payment.
[d] With penalty for late payment.
[e] For original colonists.
[f] For newcomers.
[g] Government policy.
[h] Mean hectares financed fixed at one and standard deviation at zero for loan types not based on areas.
[i] Calculated from Cr$8000 (the midpoint of the range of house prices of Cr$5000–11,000 in uncorrected cruzeiros of January 1971) + Cr$8260 (Cr$1.40/ha for land + Cr$1.20/ha for surveying charge for 100 hectares, in uncorrected cruzeiros of January 1971), corrected for inflation to January 1, 1975.
[j] Ending year set at 999 so that loans of this type are available throughout the entire run.
[k] From 1972 ACAR-PARÁ loan request forms (*cédulas*): Cr$400 of October 15, 1972.
[l] From 1974 ACAR-PARÁ loan request forms.
[m] From 8 minimum wages (totaling Cr$1342 of January 1, 1971) (Brazil, INCRA 1972a:206).
[n] Brazil, INCRA 1972a:206.
[o] Financing frequency for power saws about 0.10 (50 financed in approximately 500 colonists in 1973/74 from ACAR-PARÁ files; 6 saws financed in 61 colonists from questionnaires); financing frequency for threshing machines 0.02 (3 financed in 152 colonists).
[p] Amount of loan (not per hectare).
[q] From Cr$4500 power saw financing (uncorrected) for loan requests dated October 1973.
[r] Approximate value representing the case for most colonists.
[s] Rice seeds distributed by INCRA in January 1975 per 40 kg (1 ha).
[t] Based on 1974 ACAR-PARÁ loan request forms.
[u] From average areas of 0.9 ha for pepper and 1.98 ha for cacao in 1973 ACAR-PARÁ loan request forms (mean of 2 crops).
[v] From mean of cacao and pepper average values in 1973 ACAR-PARÁ loan request forms: Cr$5575/ha for cacao (sd = 280, n = 3), Cr75$11,920/ha for pepper (sd = 3125, n = 3).
[w] Private loan values assumed as an educated guess based on informal conversations with colonists. These loans can take the form of credit from colonist-owned shops.

First, all original colonists are granted loans for land and house purchase, and the INCRA debt for salaries and nondurables. Simulated colonists can have up to 20 loans of different types and/or years. All colonists begin as solvent. Colonists who fail to pay the principal and interest due on their loans from the Banco do Brasil or INCRA, will be classified as insolvent, and will become ineligible for further loans from these sources. An exception is the INCRA seed loans since, in my experience, these are given to all colonists regardless of solvency. Seed loans were discontinued by INCRA in the 1975–76 agricultural year. If late payments are made, a higher rate of interest is charged for many loan types. Monetary correction to adjust debts for inflation can also be included for any loan type, but the values used for this parameter make all simulated loans without monetary correction, thus giving a substantial government subsidy to the colonists. Assigning loans without monetary correction is based on consultation with several Banco do Brasil officials in Altamira, the statements of ACAR-PARÁ (now EMBRA-TER) extension agents, and the loan agreements as signed by the colonists.

As a part of the decision process for land-use allocations, the number of patches financed must be determined for each of the relevant operations. In considering an operation (such as felling) which can be aided by a loan, a check is made to see if the colonist is eligible for the loan. If eligible, a determination is made of whether the loan is granted, based on the probability of financing from table A.9. These probabilities are for colonists who both have a loan approved and pick the money up at the bank, since the slowness of the government bureaucracy often (usually) results in the loans not being liberated until after the season for the agricultural operation has passed on the Transamazon Highway.

If a loan is actually granted, then appropriate adjustments must be made to the simulated colonist's capital and labor supplies. Along with the input of capital from the loan, there is also a heavy cost to the colonist both in money and time spent on arranging for the loan. These costs are only levied in the simulation for colonists who actually receive loans, an optimistic assumption since many colonists in reality expend time and money on the initial steps of acquiring loans, but subsequently fail to get them either because the loans are not approved or are not picked up at the bank following delays. The cash cost of trips to the bank to pay the loan is not included at this point, but is subtracted later at the time of loan payment in the product allocation sector.

For loan types which are granted on a per-hectare basis, the amount of the loan added for each financed patch is calculated by multiplying the size of a patch by the amount per hectare. This is done for each patch financed. For one-time loans such as those for land purchase, the amounts are generated from the mean and standard deviation shown for the total loan amounts in table A.9. These calculations are only done once. In the case of INCRA seed loans, the number of patches financed for the crop is calculated from the areas being financed for Banco do Brasil *custeio* loans for the crop in question (areas financed in general, not the area specific to each individual colonist, who may be insolvent and still receive seed loans).

The costs to the simulated colonist for obtaining a loan is only levied once, the cost being the same whether a large or a small area is financed. Loans for settlement (house and land, INCRA debt, and durable goods) and private loans have no labor or cash costs. For other types of loans the cost is deducted from the amount of the loan. This is done in fractions as each patch is considered (the financing cost divided by the number of patches tentatively financed) to prevent the colonist's capital from becoming negative. Since programming devices such as this have been included, financing is considered in the land-use allocation procedure in determining if a colonist is capable of allotting each additional patch. Financing is considered on a tentative basis, and then all resulting changes are undone if, in the end, it is determined that the colonist is not capable of making the allotment being considered.

The value used for the mean cash cost of financing is Cr75$1041. This is the midpoint of the range of financing costs of Cr$595-1050 given by Moran (1976:86), assuming that his figures refer to cruzeiros of June 15, 1974. This is a somewhat optimistic figure, since the agrovila studied by Moran is only 23 km from the town of Altamira, whereas the intensive study area for the present carrying capacity study is centered on an agrovila 50 km from the town (with transportation costs to Altamira approximately twice as high, but transportation costs to the ACAR-PARÁ (now EMBRATER) offices in Agrópolis Brasil Novo about the same). The figure is corrected to cruzeiros of January 1, 1975.[1]

The standard deviation of the financing cash cost is estimated from the range values given by Moran (1976:86). The maximum and minimum values, corrected to cruzeiros of January 1, 1975, were used to approximate a standard deviation by the method of Snedecor and Cochran (1967: 40), assuming that Moran's sample size is 25 colonists based on other data presented in the study referred to as a "50% sample" of an agrovila (agrovilas normally house about 50 families). The value for the standard deviation thus obtained is Cr75$148. Cost of financing is generated in the simulation from the mean and standard deviation for this cost once per year for each colonist attempting to obtain financing.

Labor loss in days for arranging financing is generated in a similar way. The value for the mean labor loss used is twenty-four days, the midpoint of the range of values given by Moran (1976:86).[2] The standard deviation was approximated using the multipliers of Snedecor and Cochran (1967:40), assuming a sample size of 25 colonists. The standard deviation value obtained in this way is 3.3 days. The labor loss must also be apportioned to the various months of the year. The proportions of financing labor loss falling in each calendar month used were as follows: January: 0.11, February: 0.00, March: 0.00, April: 0.00, May: 0.11, June: 0.08, July: 0.08, August: 0.29, September: 0.11, October: 0.11, November: 0.11, and December: 0.00. These proportions are calculated from the days for each of the different financing operations, using the midpoints of the ranges given by Moran (1976:86). Operations are assigned to months as follows: making plan (July); confirmation of approval (August); receiving of six payments (equally divided between August, September, October, November, January, and May); and payment of interest on loan (June).

The proceeds from the loans are allocated in the simulation according to loan type. Loans for capital goods go directly to capital goods in the simulation, as is required by the bank but sometimes is circumvented in actual fact. Private loans go to consumption cash, INCRA debt loans are not added to any of the cash types, since colonists are assumed to feed themselves during the first year on INCRA salaries. The money from loans for houses and land never passes through the simulated colonist's hands, as is also the case with capital good loans and INCRA seed loans. Other loan types are added to the colonist's capital.

LAND CLEARING

Land clearing decisions are a part of the land-use allocation process. The patches of land available for allocation in the lot are "cleared," or prepared for planting, in an order guided by the preference of the simulated colonist for the different categories of land available for clearing.

In choosing the first patch to be cleared for a given lot and year, the first step is the determination of which patches in the lot are available for clearing. Those unavailable include patches which are planted in perennial crops or pasture. Those patches which are in manioc (sweet or bitter) at the end of the previous agricultural year are also unavailable, since the growth period of manioc is greater than one year. Patches are also considered unavailable for clearing if they have been under continuous cultivation for the maximum number of years permitted before weeds make planting a new crop impossible without an intervening fallow period. Here "continuous cultivation" includes annual crops, perennial crops, pasture, and bare and weeds (less than 240 days old). The intervening fallow period will be at least to the "second growth" (over 240 days) stage. The maximum number of years permitted in continuous cultivation is an input parameter, a value of two years being used in accord with the usual practice on the Transamazon Highway.

Each of the patches considered "available" for clearing is assigned a clearing class category, and a tally is kept of the number of patches available in each category. The categories are: 1) weeds or bare at the end of the previous agricultural year (less than 240 days); 2) second growth less than 2 years old; 3) second growth 2–3 years old; 4) second growth 4–6 years old; 5) second growth 7–11 years old; 6) second growth 12–16 years old; 7) second growth 17–20 years old; 8) second growth over 20 years old; and 9) virgin forest (forest not previously cleared by colonists).

Next, the number of patches financed for clearing of virgin forest is determined. This is done according to the procedure previously described for financing, based on the simulated colonist's eligibility for loans, areas financed in government credit programs, etc.

The first clearing done is for financed virgin felling, until the number of patches financed for virgin clearing is reached, or until the simulated colonist's resources are exhausted, whichever occurs first. Starting with a randomly selected patch, the patches in the lot are examined until a virgin patch is encountered. Appropriate adjustments are made in the simulated colonist's capital, depending on the financing terms in force in the year in question.

For patches not influenced by virgin clearing financing, the clearing category, or age category, of the patch to be cleared is chosen (among categories for which there are some patches available in the lot), based on clearing probabilities. The clearing classes represent the probabilities of clearing some of this class of land for uses other than pasture, given the existence of both virgin land and the category in question in the lot. These probabilities only apply to fields under cultivation less than two years. The probabilities are calculated as proportions of lot-years where land of the category is available in which it is cleared (Fearnside 1984g). The probabilities used are: 1) bare and weeds: 0.800 (N=60); 2) 8 months–2 years: 0.750 (N=28); 3) 2–3 years: 0.385 (N=13); 4, 5, 6, 7) 4–20 years: all assumed to be 0.385 (the same as category 3); 8) over 20 years: assumed to be 1.0 (the same as category 9); 9) virgin: 1.0. These values for clearing probabilities are input parameters used to represent the actual behavior of colonists on the Transamazon Highway. It is possible to examine the effects of hypothetical changes in this behavior, such as the maintenance of fallow periods of a certain length, by altering the clearing probabilities to make probabilities for age classes below the length of the fallow period equal to zero, and those above it equal to one.

Once the category to be cleared has been identified, a patch is selected and cleared in this category. This process continues as the land use allocation procedure progresses, and stops when land, labor, or capital resources for further allocations have been exhausted.

SUBSISTENCE AREA ALLOWANCE

Colonist Learning and Allocation Behavior

Subsistence demand is determined for the simulated colonist by calculating the number of hectares of subsistence crops needed to supply subsistence quantities of these crops, including a margin for protection against poor yields. The program contains a feature which allows colonists to learn from the experience of all colonists in the area in estimating how much land will be necessary to supply subsistence needs, including the margin for poor yields. This feature can also be disabled by the program user if so desired. To make the learning adjustments, the mean and standard deviation of crop yields must be computed.

In the first year (or in runs without the learning feature) the expected yield mean and standard deviation which have been entered as program parameters are used. These starting values are based on official publications from the colonization agency, presumably corresponding to what the colonists were told to expect when they arrived. The expected yield for rice is 1,500 kg/ha (Brazil, INCRA 1972a:195); for maize is 1,500 kg/ha (p. 196); for *Phaseolus* beans is 800 kg/ha (p. 197); for V*igna* cowpeas (also called "beans" in Portuguese) it is assumed to be the same 800 kg/ha; for bitter manioc it is 6667 kg flour/ha/year of growth (p. 196); and for sweet manioc it is assumed to be the same 6667 kg flour/ha/year of growth. The standard deviations are zero for all crops, corresponding to no variation in yields. In subsequent years, in runs with the learning feature, the mean and standard deviation are adjusted in accord with the experience of the simulated colonists.

Subsistence Margin Against Poor Yields

The hectares needed for each subsistence crop are calculated by multiplying the subsistence need for the product (calculated earlier) by the family size for the lot, adding this to the product of the "z" statistic of acceptable colonist risk and the expected standard deviation for yields of the crop, and then correcting the result for expected spoilage by multiplying by the sum of the expected spoilage proportion plus one, and dividing by the mean expected yield for the crop. The value for the "z" statistic used to calculate the safety margin protecting the colonist against failure is calculated from an official statement on acceptable risk. It is hoped that eventually better information reflecting the colonists' own point of view will be available on this hard-to-quantify item. The value used was calculated from the RADAMBRASIL Project's classification of soils in the area as having "good production" if they are "showing light risks of damage or crop failure due to irregularity in the distribution of rainfall, with a probability of occurrence of once in a period of more than five years" (Brazil, RADAMBRASIL 1974 5:III/117). The period of five years average to a failure corresponds to a probability of 0.1294 of failing per year, which corresponds to a "z" statistic value of 1.13.

In the case of "beans," where there are two types (*Phaseolus* beans and *Vigna* cowpeas), and manioc, where there are also two types (sweet and bitter), the "bean" or manioc type planted for any given year is selected and the expected yield corresponding to this type is used in the subsistence area calculations. A given colonist only uses one of these types in the simulation. The *Vigna* type probability (the probability that a colonist will plant *Vigna* rather than *Phaseolus* in a given year given that one of the two is planted) is 0.221, which is based on the proportions in the "bean" fields studied (120 *Phaseolus* fields and 34 *Vigna* fields). The sweet manioc type probability (the probability that a colonist will plant sweet as opposed to bitter manioc in a given year given that one of the two is planted) is 0.169, which is based on the proportion of the sweet variety in manioc fields studied (13 sweet vs. 64 bitter).

For all of these crops (rice, maize, "beans," and manioc) the subsistence area required is calculated as the product of the subsistence kilograms of the product per capita times the family size for the lot, plus the product of the "z" statistic for acceptable failure and the expected standard deviation in yields for the crop. This quantity is then multiplied by a correction factor for expected spoilage, which is computed by adding the expected spoilage proportion to one and then dividing the quantity by the expected mean yield per hectare for the crop.

Modeling Crop Allocations

Interplanting Decisions

Once a decision has been reached to allocate a given patch to a given crop, the land use must be determined, including interplanting with other crops. Interplanting decisions are made based on the probabilities of given combinations appearing, within the constraints of the twenty-four land uses included in the simulation. The original allocation decisions are made on the basis of single crops, and the interplanted crops are added to this principal crop after the allocation decision has been made.

The probability of rice being planted alone is 0.43 ($N = 303$ rice fields). The probability of rice being interplanted with maize (only, not maize + any other crop) is 0.41 ($N = 303$ rice fields). The probability of rice interplanted with bitter manioc is 0.07. The rice with bitter manioc probability is calculated from the overall probability of rice with any kind of manioc of 0.083 (25 of 303 rice fields) and the proportion of bitter manioc in manioc fields of any type or interplanting combination, which is 0.831 (64 bitter of 77 total manioc fields). Similarly, the probability used for rice interplanted with sweet manioc is 0.01, calculated from the 0.083 overall rice with manioc probability and the proportion of sweet manioc overall of 0.169. The probability of rice with both maize and bitter manioc used was 0.04, which is calculated from the overall rice with maize and manioc probability of 0.046 (14 of 303 rice fields), and the proportion of bitter manioc overall of 0.831. The probability of rice with both maize and sweet manioc is 0.01, based on the overall probability of rice with maize and manioc of 0.046 and the sweet manioc proportion overall of 0.169. The probability of rice interplanted with pasture is 0.03, which is based on the probability of rice with pasture of 0.026 (8 of 303 rice fields) and the probability of rice with pasture and maize of 0.003 (1 of 303 rice fields, a combination not included explicitly in the simulation).

The probability of *Phaseolus* being planted alone (as opposed to planted with green maize; dry maize is not counted as an interplanted crop) is 0.90 ($N = 120$ *Phaseolus* fields). This figure for *Phaseolus* alone actually includes some other combinations not explicitly included in the simulation: 0.808 was *Phaseolus* alone, 0.067 was *Phaseolus* with manioc, 0.008 was *Phaseolus* with dry maize, and 0.075 was *Phaseolus* with other crops. This is the only *Phaseolus* parameter used; the value used is less than one, as in the case of the sum of the probabilities of rice combinations. The remainder (0.10) represents the probability of *Phaseolus* being interplanted with green maize. This is based on a probability of 0.092 for *Phaseolus* with green maize only (11 out of 120 *Phaseolus* fields) and a probability of 0.008 for *Phaseolus* with both green maize and manioc (1 out of 120 *Phaseolus* fields, a combination not explicitly included in the simulation).

The probability of V*igna* "alone" is 0.32 (N = 34 V*igna* fields). This is the probability of V*igna* being planted either alone or with crops other than maize, either green or dry. The probabilities for other combinations are 0.147 for V*igna* with manioc and 0.147 for V*igna* with other crops. Since the V*igna* "alone" probability is the only V*igna* interplanting probability used as an input parameter, the difference between this and 1.000 represents the probability of V*igna* being interplanted with maize, either dry or green. This probability is 0.68, which is based on the probabilities of 0.206 for V*igna* with dry maize, 0.235 for V*igna* with green maize, 0.000 for V*igna* with dry maize and manioc, and 0.235 for V*igna* with green maize and manioc (N = 34 V*igna* fields).

The probability of bitter manioc alone or with crops other than rice or maize is 0.62 (N = 64 bitter manioc fields), and the probability of bitter manioc with maize is 0.08 (5 of 64 bitter manioc fields). These probabilities do not sum to one, the remaining probability of 0.30 representing the probability of bitter manioc with both rice and maize (19 of 64 bitter manioc fields).

The probability of sweet manioc being planted alone or with crops other than rice or maize is 0.91 (10 of 11 sweet manioc fields). The probability of sweet manioc interplanted with rice is 0.00 (0 of 11 sweet manioc fields), and the probability of sweet manioc interplanted with maize is also 0.00 (0 of 11 sweet manioc fields). These probabilities do not sum to one, the remaining probability of 0.09 (1 of 11 sweet manioc fields) representing sweet manioc with both rice and maize.

Interplanting decisions are not included separately for maize to avoid duplication with the interplanting decisions from other crops. Allocations which are explicitly to be made with maize as the principal crop are therefore assigned to maize alone, and all maize which is interplanted is the result of assignments with other crops as the principal crop.

Allocations for cacao, pepper, pasture without animals, and pasture with animals are made directly without interplanting.

Labor and Capital Constraints

The operations needed to install the proposed land use are first determined. For example, if the proposed land use is rice and the present land use is virgin forest, both the virgin clearing and the rice *custeio* (planting, weeding, and harvesting, including piling and threshing) must be checked. There are fifteen operations considered in the program: 1) virgin clearing *(derruba)*; 2) second growth clearing *(roçagem)* ("second growth" is defined as at least eight months uncultivated); 3) weed clearing *(limpa)* ("weeds" are defined as 2–8 months uncultivated); 4) rice *custeio* (including planting, weeding, and harvesting, but not second growth or weed clearing as in the case of *custeio* bank loans); 5) maize *custeio*; 6) *Phaseolus* or *Vigna* "bean" *custeio*; 7) sweet or bitter manioc *custeio* (including flour making for labor but not for capital checks); 8) cacao establishment; 9) pepper establishment; 10) pasture without animals establishment (planting, not including building fences or corrals); 11) pasture with animals establishment (including fencing and corrals); 12) cacao maintenance (not including fertilizer costs); 13) pepper maintenance (not including fertilizer costs); 14) pasture without animals maintenance; and 15) pasture with animals maintenance. The checks require parameters for total labor requirement (table A.10), male labor requirement (table A.11), and fixed costs for each operation. Labor requirements for the tasks making up each operation are summarized in table A.12, and fixed cost requirements are given in table A.13.

The first check made is for total labor feasibility without hired labor. For each month the total labor requirement, calculated from the total labor requirements shown in table A.10 for the operation in question and the size of the patch, is subtracted from the labor supply figures for the month. If the labor supply from family labor is insufficient, then the amount of supplementary hired labor is determined, along with its cost based on the cost of labor in that month. The costs of labor are based on the going rates in the 1974/75 agricultural year, which were Cr$15/day for all months except for the rice harvest period (June) and the felling period (August–September), which were Cr$20/day (with no food included in the bargain for either rate). If the capital figure for the lot is not sufficient to pay for the hired labor required, then the labor and capital sufficiency check is unsuccessful.

If total labor supplies are adequate, a check is made of male labor requirements to determine if additional hired labor is necessary before a successful labor and capital sufficiency check result can be returned. The male labor available for each month is calculated from the male family labor plus the

Table A.10. Total Labor Requirements for Agricultural Operations by Month

Operation	Jan.	Feb.	March	April	May	June	July	Aug.	Sept.	Oct.	Nov.	Dec.
Clearing												
1. Virgin	0.0	0.0	0.0	0.0	0.0	0.0	6.9	6.9	6.9	8.0	0.0	0.0
2. Second growth	0.0	0.0	0.0	0.0	0.0	0.0	0.0	0.0	10.0	10.0	4.8	0.0
3. Weed	0.0	0.0	0.0	0.0	0.0	0.0	0.0	0.0	0.0	7.0	1.7	0.0
Crops												
4. Rice[a]	6.2	3.7	3.7	0.0	0.0	14.1	4.1	0.0	0.0	0.0	0.0	0.0
5. Maize[a]	0.0	4.7	4.7	0.0	0.0	0.0	0.0	5.0	5.0	0.0	0.0	2.0
6. Beans[a]	0.0	0.0	0.0	13.4	0.0	6.2	11.2	0.0	0.0	0.0	0.0	0.0
7. Manioc[b]	0.0	20.3	20.3	20.3	20.3	0.0	0.0	0.0	0.0	0.0	20.3	20.3
Establishment												
8. Cacao	20.0	0.0	0.0	0.0	0.0	0.0	0.0	15.0	6.0	0.0	22.0	22.0
9. Pepper	27.0	13.5	1.5	4.5	12.0	0.0	46.3	46.3	10.0	30.0	24.0	0.0
10. Pasture without animals	2.3	0.0	0.0	0.0	0.0	0.0	0.0	0.0	0.0	0.0	0.0	0.0
11. Pasture with animals	0.0	0.0	0.0	0.0	0.0	0.0	4.7	4.7	0.0	0.0	0.0	0.0
Maintenance												
12. Cacao	9.0	9.0	9.0	9.0	9.0	9.0	9.0	9.0	9.0	9.0	9.0	9.0
13. Pepper	1.5	9.0	2.5	1.5	0.0	11.5	1.5	9.0	2.5	0.0	9.0	2.5
14. Pasture without animals	0.0	0.0	0.0	0.0	0.0	0.0	1.0	1.0	0.0	0.0	0.0	0.0
15. Pasture with animals	0.6	0.6	0.6	0.6	0.6	0.6	4.1	4.1	0.6	0.6	0.6	0.6

SOURCE: Fearnside 1980b.
NOTE: Total labor requirement (regardless of age and sex) in man-day equivalents/ha (see text for justifications).
[a] Planting, weeding and harvesting.
[b] Planting, weeding, harvesting and flour (*farinha*) making.

Table A.11. Male Labor Requirements for Agricultural Operations by Month

Operation	Jan.	Feb.	March	April	May	June	July	Aug.	Sept.	Oct.	Nov.	Dec.
Clearing												
1. Virgin	0.0	0.0	0.0	0.0	0.0	0.0	6.9	6.9	6.9	8.0	0.0	0.0
2. Second growth	0.0	0.0	0.0	0.0	0.0	0.0	0.0	0.0	10.0	10.0	4.8	0.0
3. Weed	0.0	0.0	0.0	0.0	0.0	0.0	0.0	0.0	0.0	7.0	1.7	0.0
Crops												
4. Rice[a]	0.0	0.0	0.0	0.0	0.0	14.1	4.1	0.0	0.0	0.0	0.0	0.0
5. Maize[b]	0.0	0.0	0.0	0.0	0.0	0.0	0.0	2.5	2.5	0.0	0.0	2.0
6. Beans (*Phaseolus* or *Vigna*)[a]	0.0	0.0	0.0	13.4	0.0	0.0	5.0	0.0	0.0	0.0	0.0	0.0
7. Manioc (sweet or bitter)[b]	0.0	10.8	10.8	10.8	10.8	0.0	0.0	0.0	0.0	0.0	10.8	10.8
Establishment												
8. Cacao	20.0	0.0	0.0	0.0	0.0	0.0	0.0	9.0	6.0	0.0	22.0	22.0
9. Pepper	27.0	0.0	1.5	0.0	0.0	0.0	46.3	46.3	10.0	30.0	24.0	0.0
10. Pasture without animals	2.3	0.0	0.0	0.0	0.0	0.0	0.0	0.0	0.0	0.0	0.0	0.0
11. Pasture with animals	0.0	0.0	0.0	0.0	0.0	0.0	4.7	4.7	0.0	0.0	0.0	0.0
Maintenance												
12. Cacao	5.5	5.5	5.5	5.5	5.5	5.5	5.5	5.5	5.5	5.5	5.5	5.5
13. Pepper	1.5	0.0	0.0	1.5	0.0	10.0	1.5	0.0	0.0	0.0	0.0	0.0
14. Pasture without animals	0.0	0.0	0.0	0.0	0.0	0.0	1.0	1.0	0.0	0.0	0.0	0.0
15. Pasture with animals	0.6	0.6	0.6	0.6	0.6	0.6	4.1	4.1	0.6	0.6	0.6	0.6

SOURCE: Fearnside 1980b.
NOTE: Requirement for male labor (adult men at least 18 years of age) in man-days/ha. See text for justifications.
[a] Planting, weeding, and harvesting.
[b] Planting, weeding, harvesting, and flour (*farinha*) making.

Table A.12. Labor Requirements for Agricultural Operations

Operation	Task	Months	Mean Total Labor Requirement (man-days/ha)	SD	N	Source	Male Labor (% of total)	Notes
Clearing								
1. Virgin	Underclearing	July, Aug.	11.45	7.04	21	field data	100	
	Felling	Aug., Sept.	9.34	3.65	12	field data	100	
	Burning	Oct.	1.69	3.80	15	field data	100	
	Coivara	Oct.	6.26	8.86	200	field data	100	Piling up unburned material for a second burn
2. Second growth	Cutting	Sept., Oct.	20.0	—	1	field data	100	
	Coivara	Nov.	4.79	7.06	40	field data	100	
3. Weed	Cutting	Oct.	7.00	5.18	6	field data	100	Preparation for planting rice
	Coivara	Nov.	1.66	2.32	12	field data	100	
Crops								
4. Rice	Planting	Jan.	6.23	6.26	13	field data	100	
	Weeding	Feb., March	7.40	5.40	207	field data	100	
	Harvesting	June	14.08	10.14	'2	field data	100	
	Threshing	July	4.07	1.26	4	field data	100	For hand threshing. Machine threshing requires 2 man-days/ha (Smith 1976b:158)
5. Maize	Planting	Dec.	2.0			Smith 1976b:194	100	Assumes maize is planted alone
	Weeding	Feb., March	9.93	9.43	141	field data	0	
	Harvesting	Aug.	5			Smith 1976b:194	50	
	Remove kernels	Sept.	5			Smith 1976b:194	50	
6. Beans (either *Phaseolus* or *Vigna*)	*Limpa*	April	10.9	(see note)		field data	100	*Limpa* (clearing weeds in preparation for planting) not considered a "clearing"
	Phaseolus		14.8	10.3	10	field data	100	
	Vigna		7.00	3.00	3	field data	100	

Table A.12. Labor Requirements for Agricultural Operations (*Continued*)

Operation	Task	Months	Mean Total Labor Requirement (man-days/ha)	SD	N	Source	Male Labor (% of total)	Notes
	Planting	April		(see note)	2	field data		operation in the case of beans; midpoints between the means for the two species used for requirements
	Phaseolus		2.49	1.16			0	
	Vigna		2.49		0			
	Weeding	June, July		(see note)				
	Phaseolus		12.34	7.50	63	field data	0	
	Vigna		9.54	20.07	22	field data	0	
	Harvesting	July		(see note)				
	Phaseolus		15.16	—	1	field data	100	
	Vigna		4.96		0			
			4.96					
7. Manioc (either sweet or bitter)	Planting	Feb.–May; Nov.–Dec. (all labor spread over six least busy months)	4.07	4.39	2	field data	100	All values corrected to give labor requirements per year using the mean bitter manioc growth period of 1.29 years (SD = 0.53, N = 64); flour (*farinha*) making labor from mean bitter manioc yield of 3617.7 kg flour/ha/year of growth (SD = 2002.2, N = 15) from field data, and processing figure of 37 kg flour/man-day (Smith 1976b:158)
	weeding		8.53	6.05	27	field data	0	
	harvesting		11.63	—	1	field data	100	
	flour making		97.77			field data; Smith 1976b:158	50	
Establishment								
8. Cacao	Preparing area	Dec.	12.0			All figures from Brazil, INCRA 1972a:168	100	Labor figures calculated from family labor equivalents given for family of 2.5 adult male equivalents, derived using the
	Provisional shading	Dec.	10.0				100	
	Definitive shading	Jan.	4.0				100	
	Nursery construction	Aug.	5.0				100	

	Activity	Month	Value	Source	%	Notes
	Filling plastic bags	Aug.	6.0		0	same schedule of equivalents used in the present study (table A.37)
	Planting seeds	Aug.	4.0		0	
	Care of seedlings	Sept.	6.0		0	
	Holes for seedlings	Nov.	16.0		100	
	Fertilizing	Nov.	2.0		100	
	Planting seedlings	Jan.	16.0		100	
	Liming	Nov.	4.0		100	
9. Black pepper	Planting cuttings and fertilizing	Jan.	27.0	All figures from Brazil, INCRA 1972a:169	100	
	Weeding during establishment	Feb.	9.0		0	
	Mounding	Feb., Apr.	12.0		0	
	Treatments	Mar.	1.5		100	
	Mulching	May	12.0		0	
	Post cutting	Jul.	46.3	guess	100	20 minutes/post
	Post transport	Aug.	46.3	guess	100	20 minutes/post
	Cleaning area	Sept.	10.0	Brazil, INCRA 1972a:169	100	
	Hole digging	Oct.	30.0		100	13 minutes/hole
	Post placing	Nov.	24.0		100	10 minutes/post
10. Pasture without animals	Seed gathering	Jan.	2.3	field data	1	3 man-days/60kg sack, sufficient to sow 1.3 ha of *Panicum maximum*
11. Pasture with animals	Fence post cutting	July–Aug.	0.89	guess	100	20 minutes/post (spacing 5 m, field area = 22.7ha, N=1)
	Post transport	July–Aug.	0.89	guess	100	20 minutes/post
	Hole digging	July–Aug.	0.58	Brazil, INCRA 1972a:169	100	13 minutes/hole, value for black pepper posts
	Post placing	July–Aug.	0.46	Brazil, INCRA 1972a:169	100	For black pepper posts
	Stringing wire	July–Aug.	0.89	guess	100	20 minutes/post
	Corral construction	July–Aug.	4.76	field data	1	150 m^2 corral

Table A.12. Labor Requirements for Agricultural Operations (*Continued*)

Operation	Task	Months	Mean Total Labor Requirement (man-days/ha)	SD	N	Source	Male Labor (% of total)	Notes
Maintenance								
12. Cacao	Fertilizing	All tasks spread over entire year	16.67			Brazil, ACAR-PARÁ, Unidade Operacional Altamira VI, n.d.[b] (c. 1974)	100	Converted from value in man-days/1000 plants using the 1111 trees/ha recommended planting density
	Spraying		13.33				100	
	Pruning		11.11				100	
	Harvesting and processing		48.89				50	
	Weding		17.78				0	
13. Pepper	Weding	Feb., June, Aug., Nov.	35			Brazil, INCRA 1972a:169	0	
	Pruning	May, June, Sept., Dec.	10				100	
	Spraying	Jan., April, July	6				100	
14. Pasture without animals	Cutting invading second growth	July, Aug.	2			guess	100	Based on low maintenance standard in area
15. Pasture with animals	Cattle tending and repairs	all 12 months	6.60		2	Fearnside and Rankin, 1973 field notes	100	From two ranches near Santarém
	Cutting invading second growth	July, Aug.	7.00	5.18	6	field data	100	Assumed equal to the cutting portion of the weed clearing operation

SOURCE: Fearnside 1980b.

Table A.13. Fixed Cash Costs for Agricultural Operations

Operation	Item	Cost (Cr75$/ha)	SD	N	Notes
Clearing					
1. Virgin	Tools	0			Cost of manual tools not believed to pose any significant limit using family labor; power saws can be substituted for labor at roughly the same cost per hectare as hired labor, and therefore are not included explicitly in the program
2. Second growth	Tools	0			
3. Weed	Tools	0			
Crops					
4. Rice	Chemicals in piles (*pilhas*) of cut rice	17.75	18.59	8	Cost of sacks for all crops is deducted from the selling price rather than being considered a fixed cost, since sacks can often be obtained on credit at harvest time
5. Maize	Chemicals, etc.	0	0	4	
6. Beans (*Phaseolus* or *Vigna*)	Chemicals in seed treatment	7.60	—	2	
7. Manioc (bitter or sweet)	Cultivation	0			Costs of processing (depreciation and operation of equipment) are deducted from selling prices in simulation—not a prerequisite for planting; on the Transamazon Highway colonists without equipment can use neighbor's *casa da farinha* in exchange for 30% of production
	Flour (*farinha*) making	0			
Establishment					
8. Cacao	Aldrin (6 kg)	74			From: Brazil, SAGRI e CEPLAC (n.d. [c. 1974]); values converted to Cr75$
	Rustic constructions	77			
	Plastic bags	82			
	Sprayer	315			
9. Pepper	Chemicals, etc.	16.38			From: Brazil, ACAR-PARÁ (1973); value converted to Cr75$; assumes recommended density of 1111 plants/ha

Table A.13. Fixed Cash Costs for Agricultural Operations (*Continued*)

Operation	Item	Cost (Cr75$/ha)	SD	N	Notes
10. Pasture without animals	Seeds, tools, etc.	0			Not considered a significant barrier to planting
11. Pasture with animals	Wire for fencing	74.34			Four-strand fence based on Altamira price: Cr$350 for 500 m roll in February 1975
	Cattle	459.00	—	1	Cr75$2500/head purchase price + Cr75$200/head for transportation, assuming stocking rate of 0.17 head/ha
	Wire for corral	154.17			
Maintenance					
12. Cacao	Chemical treatment	61.52			40 kg/ha/year of 1% BHC insecticide costing Cr$1.00/kg in 1974 (Brazil, SAGRI e CEPLAC (n.d [c. 1974]); depreciation of equipment calculated elsewhere in program
13. Pepper	Chemicals + fertilizers	995.12			Inputs for 1000 plants are: 15 kg "Cuprovit" or "Cuprosan" fungicide, 1 liter 100% "Malatol" insecticide, 12 kg "Dithane M-45" insecticide, 600 kg NPK fertilizer, 4 liters "Novapol," 3000 kg castor bean (*Ricinus communis*) cake, 500 kg dolomitic lime, and 500 kg bone meal (Brazil, ACAR-PARÁ 1973); correction made for inflation and 1111 plants/ha recommended density
14. Pasture without animals	Tools, etc.	0			Not considered a significant limitation
15. Pasture with animals	Fence repairs, medicines	300.00			Assumed

SOURCE: Fearnside 1980b.

hired labor (all hired labor is considered to be adult males), less the male labor requirement as calculated from the product of per-hectare male labor requirement and the size of a patch. If the male labor is insufficient, then the feasibility of hiring outside labor to meet this requirement is tested. If the capital figure is greater than the cost of labor times the absolute value of the male labor deficit for the month, then the labor is hired with appropriate adjustments to capital. Otherwise the check is unsuccessful. If hiring labor to meet the male labor deficit is found to be feasible, then the product of the cost of labor and man-days required is subtracted from the capital figure, and the figure for the amount of hired labor used is updated to reflect the hiring.

Fixed capital cost feasibility is the last requirement checked. For this, capital is reduced by the product of the per-hectare fixed cost requirement for the operation and month and the size of a patch. If capital becomes negative, then the check is unsuccessful. If the check is successful to this point, then a determination is made of whether a second round of checks is necessary.

If the operation which has just been checked is the first part of a two-part operation such as an annual crop on a virgin forest site, where one check must be made for the clearing operation and another for the requirements from planting through harvesting, then the operation (a clearing operation) is reassigned (as a crop operation) and the above described total labor, male labor, and fixed cost checks are performed on the new operation. If the check is unsuccessful at any point, then the unsuccessful result is indicated for the land use.

Cash Crop Choices

The probabilities of using various crops as cash crops are determined for the simulated lot using probabilities for single crops (not combinations of interplanted crops) being used as cash crops, based on frequencies observed among colonists following each of the four land-use patterns on the Transamazon Highway. The probabilities of land use employed as parameters for the program have been calculated from the data as the quotient of the cash crop hectare-years of the crop for a given land-use pattern divided by the total cash crop hectare-years of the crop for that land-use pattern. "Cash hectares" are areas in excess of figures considered reasonable for subsistence needs alone depending on the crop. These areas are for single crop assignments, not interplanted combinations. The division between subsistence and cash areas used was 0.5 ha in the case of rice, maize, *Phaseolus*, and *Vigna*; 0.2 ha in the case of bitter manioc and sweet manioc, and 0.0 ha in the case of cacao, pepper, pasture without animals, and pasture with animals. "Cash hectare-years" is the product of the cash hectares and the number of colonist-years in which the land was allocated to this crop (see table 4.5).

Modeling Soil Changes

BURN QUALITIES

Table A.14. Virgin Forest Felling-Month Distribution

	Month								
Item	May	June	July	Aug.	Sept.	Oct.	Nov.	Dec.	Total
Number	1	2	16	31	180	82	37	14	363
Percent	0.3	0.6	4.4	8.5	49.5	22.6	10.2	3.9	100

SOURCE: Fearnside n.d.(c)
NOTE: Mean days between felling and burning = 44.1 (SD = 65.3, N = 138).

Simulated virgin burns are classified as "bad" or "good" depending on whether equation A.5 or A.6 gives the higher value.

$$Y = 3.2459 \times 10^{-3} \, A - 3.5933 \times 10^{-3} \, B + 3.4928 \times 10^{-3} \, C + 7.7949 \times 10^{-2} \, D + 1.5809 \times 10^{-1} \, E + 3.8381 \times 10^{-2} \, F - 6.1617 \quad \text{Equation A.5}$$

where:

Y = bad burn discriminator
A = rain between felling and burning (mm)
B = evaporation between felling and burning (hours)
C = insolation between felling and burning (hours)
D = rain in 15 days prior to burn (mm)
E = evaporation in 15 days prior to burn (mm)
F = insolation in 15 days prior to burn (hours)
(general variance = 2.43×10^{22}, N = 76).

$$Y = 1.2662 \times 10^{-3} \, A - 5.2735 \times 10^{-5} \, B + 2.5793 \times 10^{-3} \, C + 8.8626 \times 10^{-2} \, D + 1.827 \times 10^{-2} \, E + 3.1593 \times 10^{-2} \, F - 7.5752 \quad \text{Equation A.6}$$

where:

Y = good burn discriminator
A–F = same as for equation A.5
(general variance = 2.26×10^{20}, N = 171).

Statistics for discriminant functions (equations A.5 and A.6):
Mahalanobis distance (D^2) = 0.686, F statistic = 5.89, $p<0.001$;
Equality of covariances: df = 21, 93234, F statistic = 22.47, $p<0.0001$.
Correctly predicted cases = 74%, N = 247.

Table A.15. Second Growth Cutting and Burning Month Distributions

Item	June	July	Aug.	Sept.	Oct.	Nov.	Dec.	Jan.	Total
				Cutting					
Number	1	8	8	39	24	20	10	1	111
Percent	0.9	7.2	7.2	35.1	21.6	18.0	9.0	0.9	100
				Burning					
Number	0	0	4	10	50	37	13	5	119
Percent	0	0	3.4	8.4	42.0	31.1	10.9	4.2	100

SOURCE: Fearnside n.d.(c)
NOTE: Mean days between cutting and burning = 52.6 (SD = 96.1, N = 79).

Simulated second growth burns are classified between "bad" and "good" using as discriminant functions equations A.7 and A.8:

$$Y = 4.8378 \times 10^{-4} A - 1.3939 \times 10^{-2} B + 2.9030 \times 10^{-3} C - 1.3692 \times 10^{-1} \quad \text{Equation A.7}$$

where:

Y = bad burn discriminator
A = rain between cutting and burning (mm)
B = evaporation between cutting and burning (mm)
C = insolation between cutting and burning (hours)
(general variance = 9.92×10^{11}, N = 31).

$$Y = -3.3761 \times 10^{-3} A - 2.0641 \times 10^{-2} B + 6.0930 \times 10^{-4} C - 1.0033 \quad \text{Equation A.8}$$

where:

Y = good burn discriminator
A–C = same as in equation A.7
(general variance = 1.29×10^{13}, N = 23).

Statistics for discriminant functions (equations A.7 and A.8):
Mahalanobis distance (D^2) = 0.566, F statistic = 2.39, p = 0.8;
Equality of covariances: df = 6, 15499, F statistic = 5.78, $p<0.0001$.
Correctly predicted cases = 65%, N = 54.

SOIL CHANGES FROM BURNING

Virgin Forest Burns

Changes in soil characters with burning can be modeled with the following set of equations. In cases where the quality of the burn is a significant factor, this is represented in multiple regression equations through the use of "dummy" variables (Draper and Smith 1966:134–41). These variables represent the effects of "good" and "bad" burns by taking on values of −1.0 and 1.0 respectively.

CHANGES IN pH

Predicting pH changes must be done separately depending on the range of the initial pH. The pH changes for cases with an initial pH of less than 4.0 are given by

$$Y = 1.538 - 0.266\ A - 0.230\ B \qquad \text{Equation A.9}$$

where:

Y = pH change
A = virgin burn quality dummy variable
  (+1 if bad; −1 if good)
B = initial Al^{+++} (meq/100g)
($p<0.0001$, r = 0.69, SE = 0.609, N = 87).

In cases where the initial pH is from 4.0 to 5.0, the pH change with virgin burning is given by

$$Y = 1.888 - 3.11 \times 10^{-2}\ A - 6.68 \times 10^{-2}\ B \qquad \text{Equation A.10}$$

where:

Y = pH change
A = initial total clay (% dry weight)
B = predicted erosion per year (mm)
($p<0.0001$, r = 0.54, SE = 0.714, N = 67).

In cases with initial pH greater than 5.0, pH change with virgin burns is given by:

$$Y = 5.207 - 0.180\,A - 0.814\,B - 6.09 \times 10^{-4}\,C \qquad \text{Equation A.11}$$

where:

Y = pH change
A = virgin burn quality dummy variable (+1 for bad burns; −1 for good burns)
B = initial pH
C = days in annual crops
($p < 0.0001$, $r = 0.56$, SE = 0.717, N = 180).

CHANGES IN ALUMINUM

Changes in aluminum with virgin burns are given by:

$$Y = 0.295 - 0.222\,A - 0.224\,B \qquad \text{Equation A.12}$$

where:

Y = aluminum change (meq/100g)
A = initial Al^{+++} (meq/100g)
B = virgin burn quality dummy variable (+1 if bad; −1 if good)
($p < 0.0001$, $r = 0.37$, SE = 1.489, N = 299).

CHANGES IN PHOSPHORUS

Phosphorus changes with virgin burns are given by:

$$Y = -0.778 + 0.677\,A - 0.357\,B \qquad \text{Equation A.13}$$

where:

Y = phosphorus change (ppm)
A = predicted phosphorus change from effects other than burning (equations A.23, A.24, and A.25)
B = virgin burn quality dummy variable (+1 if bad; −1 if good)
($p < 0.0001$, $r = 0.46$, SE = 3.255, N = 473).

CHANGES IN NITROGEN

Nitrogen changes with virgin burns are given by:

$$Y = -5.80 \times 10^{-2} - 0.654\,A + 4.89 \times 10^{-2}\,B + 2.63 \times 10^{-2}\,C$$

where: \hfill Equation A.14

Y = nitrogen change (% dry weight)
A = initial nitrogen (% dry weight)
B = initial carbon (% dry weight)
C = initial pH
$(p<0.01, \text{r}=0.46, \text{SE}=5.88 \times 10^{-2}, \text{N}=52)$.

CHANGES IN CARBON

Carbon changes are calculated using the same relationship employed for carbon changes in unburned and other fields (equation A.27). The same relationship also holds for second growth and weed burns.

Second Growth Burns

CHANGES IN pH

Changes in pH from burning second growth (fields at least eight months fallow) are given by:

$$Y = 3.4817 - 0.22603 \text{ A} - 0.23129 \text{ B} - 0.51758 \text{ C} - 3.2683 \times 10^{-4} \text{ D} \quad \text{Equation A.15}$$

where:

Y = pH change
A = initial Al^{+++} (meq/100g)
B = aluminum change (meq/100g)
C = initial pH
D = days in annual crops
$(p=0.0001, \text{r}=0.51, \text{SE}=0.646, \text{N}=91)$.

CHANGES IN ALUMINUM

Aluminum changes under second growth burns are given by:

$$Y = 0.16551 - 0.26687 \text{ A} \quad \text{Equation A.16}$$

where:

Y = aluminum change (meq/100g)
A = initial Al^{+++} (meq/100g)
$(p<0.05, \text{r}=0.37, \text{SE}=1.53, \text{N}=42)$.

CHANGES IN PHOSPHORUS

Phosphorus changes under second growth burns are given by:

$$Y = -1.5170 + 0.74065 \text{ A} - 0.83055 \text{ B} \quad \text{Equation A.17}$$

where:

Y = phosphorus change (ppm)
A = predicted phosphorus change from regression for effects other than burning (equations A.23, A.24, and A.25)
B = second growth burn quality dummy variable
 (+1 if bad; −1 if good)
$(p<0.0001, r=0.56, \text{SE}=2.620, \text{N}=78)$.

Weed Burns

The effects of weed burns, defined as fields less than eight months fallow, can be predicted from the following equations. Weed burn quality does not have a significant effect on any of these soil fertility changes.

CHANGES IN pH

Changes in pH with weed burns are given by:

$$Y = 2.9749 - 0.16504\ A - 0.51659\ B \qquad \text{Equation A.18}$$

where:

Y = pH change
A = initial pH
B = initial Al^{+++} (meq/100g)
$(p<0.001, r=0.51, \text{SE}=0.74525, \text{N}=62)$.

CHANGES IN ALUMINUM

Aluminum changes with weed burns are given by:

$$Y = 0.55043 - 0.39232\ B \qquad \text{Equation A.19}$$

where:

Y = aluminum change (meq/100g)
A = initial aluminum (meq/100g)
$(p<0.05, r=0.43, \text{SE}=1.8503, \text{N}=28)$.

CHANGES IN PHOSPHORUS

The probability that there will be no change in phosphorus with weed burns is 0.262 (N=61). For non-zero changes, the change can be predicted from:

$$Y = 3.9375 + 1.2668\ A \qquad \text{Equation A.20}$$

where:

Y = phosphorus change (ppm)
A = predicted phosphorus change per year from regression for effects other than burning (equations A.23, A.24, and A.25)
($p<0.0001$, r = 0.50, SE = 5.0023, N = 45).

SOIL CHANGES WITHOUT BURNING

Changes in soil under land uses other than pasture resulting from processes other than burning are represented by the following equations. Changes are for periods of one year.
Changes in pH are given by:

$$Y = 1.8594 - 0.4187\ A \qquad \text{Equation A.21}$$

where:

Y = pH change
A = pH of field at beginning of year
($p<0.0001$, r = 0.48, SE = 0.8418, N = 270).

Changes in aluminum ion concentration are given by:

$$Y = 4.8516 \times 10^{-6} - 1.5033 \times 10^{-5} - 1.5132 \times 10^{-4}\ B \qquad \text{Equation A.22}$$

where:

Y = Al^{+++} change (meq/100g)
A = Al^{+++} of field at beginning of year (meq/100g)
B = pH change
($p<0.0001$, r = 0.53, SE = 1.4465×10^{-4}, N = 97).

The above equation excludes fields with aluminum equal to zero in either the "before" or "after" condition.

For predicting phosphorus change, separate relationships are needed depending on the range of the initial phosphorus level. For initial phosphorus levels of 1 ppm total phosphorus, the probability of zero change is 0.648 (N = 108). Non-zero changes are given by:

$$Y = -3.4733 + 1.4143\ A \qquad \text{Equation A.23}$$

where:

Y = phosphorus change (ppm)
A = pH change of field at beginning of year
($p<0.001$, r = 0.58, SE = 1.9841, N = 34).

Phosphorus changes (total P) in cases where the initial phosphorus level is in the 2–9 ppm range are given by:

$$Y = 2.1671 + 0.97151\ A - 1.0405\ B + 2.22395 + 10^{-2}\ C$$
Equation A.24

where:

Y = phosphorus (ppm)
A = pH change
B = phosphorus of field at beginning of year (ppm)
C = proportion of time field bare or in annual crops
($p < 0.0001$, $r = 0.69$, SE $= 2.5049$, N $= 139$).

Phosphorus changes in cases where the initial phosphorus level is 10 ppm or over are given by:

$$Y = 6.8086 - 1.4363\ A \qquad \text{Equation A.25}$$

where:

Y = phosphorus change (ppm)
A = phosphorus level at beginning of year (ppm)
($p = 0.0001$, $r = 0.86$, SE $= 6.5830$, N $= 13$).

Nitrogen change (total N), excluding fields fallow for three or more years, is given by:

$$Y = -7.0371 \times 10^{-2} - 0.77136\ A + 9.1644 \times 10^{-2}\ B \\ - 3.3756 \times 10^{-5}\ C - 1.2951 \times 10^{-2}\ D + 4.1763 \times 10^{-5}\ E$$
Equation A.26

where:

Y = nitrogen change (% dry weight)
A = initial nitrogen (% dry weight)
B = initial carbon (% dry weight)
C = days in annual crops
D = initial pH
E = days fallow
($p < 0.0001$, $r = 0.77$, SE $= 3.6064 \times 10^{-2}$, N $= 114$).

Carbon change is given by equation A.27. In this case, rather than excluding all pasture, fields with pasture for more than 25% of the comparison interval were excluded.

$$Y = 0.1156 + 4.3151\ A - 0.52549\ B + 3.8721\ C \qquad \text{Equation A.27}$$

where:

Y = carbon change (% dry weight)
A = nitrogen change (% dry weight)
B = initial carbon (% dry weight)
C = initial nitrogen (% dry weight)
($p<0.0001$, $r=0.69$, SE$=0.21106$, $N=193$).

The carbon change equation is used for all fields, including burned fields and pasture.

Soil Changes Under Pasture

Soil changes under pasture are simulated using the following equations. Changes include the sign of the change. Phosphorus change under pasture is given by:

$$Y = 1.28 - 0.622\, A \qquad \text{Equation A.28}$$

where:

Y = phosphorus change (ppm)
A = initial phosphorus (ppm)
($p<0.05$, $r=0.76$, SE$=1.23$, $N=10$).

Nitrogen change under pasture is given by:

$$Y = 0.094 - 0.691\, A \qquad \text{Equation A.29}$$

where:

Y = nitrogen change (% dry weight)
A = initial nitrogen (% dry weight)
($p<0.05$, $r=0.73$, SE$=0.059$, $N=10$).

Carbon change under pasture is given by equation A.30. It should be noted that an uneven distribution of initial carbon values makes this relationship unreliable.

$$Y = 0.853 - 0.655\, A \qquad \text{Equation A.30}$$

where:

Y = carbon change (% dry weight)
A = initial carbon (% dry weight)
($p<0.05$, $r=0.76$, SE$=0.383$, $N=10$).

Changes in pH are given by:

$$Y = 3.139 - 0.875\ A - 0.547\ B \qquad \text{Equation A.31}$$

where:

Y = pH change
A = initial pH
B = inverse of years in the comparison interval
($p<0.01$, r = 0.86, SE = 0.250, N = 10).

Aluminum is calculated using the relation for initial soil quality (equation A.1).

FERTILIZERS AND LIME

The simulated colonist's labor and capital supplies must be adjusted for maintenance of perennial crops and pasture already established in the lot. Established crops are always assumed to be maintained (after satisfying subsistence needs) regardless of labor and capital "feasibility."

The family (total) labor, male labor, and capital required for the maintenance operations is determined for each month of the year. The deficit in either of the labor requirements, if any, is calculated, and this is met with hired labor (with appropriate deductions from capital supplies) if possible.

For cacao and pepper, fertilizing and liming for maintenance after the year of planting is determined separately in the simulation, since not all real colonists fertilize (although all simulated colonists can be made to fertilize by an appropriate setting of the probabilities of fertilization). A decision for fertilizing of cacao or pepper is made once per lot and year, at the time of the first maintenance operation for the crop involved. The value used for the probability of fertilizing pepper is 0.539 (N = 26 colonists in government records: 20 cases from Brazil, ACAR-PARÁ 1974a). The probability used for fertilizing cacao is 0.250, a compromise between the government-assumed probability of one and the probability of zero indicated from field data (N = 4).

If a perennial crop is fertilized, it is assumed that government-recommended dosages are followed. If pepper is to be fertilized, the doses are calculated based on the levels of the nutrients in the soil in the patch and the age of the pepper plants. The dosages used for the various categories are shown in table A.16. Liming is only done in the first year of the crop, also following the dosage of table A.16.

If cacao is fertilized, a similar schedule of dosages for fertilizers and lime

Table A.16. Government Fertilizer Recommendations for Pepper

Initial Soil Analysis	Fertilizer Active Ingredient	Dose of Active Ingredient (kg/ha) Pepper age (years)			
		1	2	3	4 or more
$P \leq 10$ ppm	P_2O_5	70	100	150	300
$P > 10$ ppm	P_2O_5	30	40	50	100
$K \leq 45$ ppm	K_2O	60	80	100	200
$K > 45$ ppm	K_2O	0	0	25	50
N (all levels)	N	40	60	80	100
C (all levels)	cotton cake[a]	2222	2222	2222	2222
$Al^{+++} \leq 0.3$ meq/100g and: $Ca^{++} + Mg^{++}$ <4 meq/100g	dolomitic lime	0	0	140	280
$Al^{+++} \leq 0.3$ meq/100g and: $Ca^{++} + Mg^{++} \leq 4$ meq/100g	dolomitic lime	122	140	280	560
$Al^{+++} \geq 0.3$ meq/100g	dolomitic lime	[b]	0	0	0

SOURCE: Fearnside 1980a; Brazil, IPEAN 1966.
[a] 5.71 kg manure is equivalent to 1 kg cotton cake (de Albuquerque and Condurú 1971:110).
[b] 2000 kg/ha lime per unit of Al^{+++} expressed in meq/100g.

is followed depending on soil quality and plant age. This is given in table A.17. Liming is also done only in the first year of the crop.

The cost of the fertilizers and lime used is deducted from the simulated colonist's capital. Prices of fertilizers and lime are given in table A.18. Fertilizing is done before liming; if a colonist does not have sufficient capital to fertilize a patch, then neither fertilizing nor liming is done. If a colonist has fertilized, but does not have enough money to lime a patch, then liming is not done.

The changes in soil nutrients from fertilizing and liming are made when a patch is fertilized or limed. Changes in pH are calculated from:

$$Y = 0.009 + 4.05 \times 10^{-4} A \qquad \text{Equation A.32}$$

where:

Y = the pH change
A = the lime dose in kg/ha dolomitic lime
($p < 0.0001$, $r = 0.99$, N = 10).

The data for the regression in equation A.32 come from a liming experiment in the Zona Bragantina of Pará, where the soil is a yellow latosol (UL-

Table A.17. Government Fertilizer Recommendations for Cacao

Initial Soil Analysis	Fertilizer Active Ingredient	Cacao Age (years)			Percent Active Ingredient[a]	Fertilizer
		1	2	3 or more		
P $\leq$ 10 ppm	P_2O_5	25	50	100	48	Triple superphosphate
P > 10 ppm	P_2O_5	0	10	25	48	Triple superphosphate
K $\leq$ 45 ppm	K_2O	25	50	200	60	Potassium chloride
K > 45 ppm	K_2O	0	10	50	60	Potassium chloride
N all levels	N	10	20	50	20	Ammonium sulfate
$Al^{+++} \geq 0.2$ meq/100g	dolomitic lime	2000 kg/ha per unit Al^{+++} expressed in meq/100g			100	Dolomitic lime

SOURCE: Brazil, IPEAN 1966.
NOTE: kg/ha active ingredient.
[a]Cruz et al. 1971:6.

Table A.18. Prices of Fertilizers and Lime in Altamira

Item	Date	Price at Date	Cr75$/kg[a]	Active Ingredient	Percent Active Ingredient	Cr75$/kg Active Ingredient
Superphosphate (triple)	April 10, 1976	4.70[b]	2.72	P_2O_5	46[c]	5.40
Urea	April 10, 1976	4.60[b]	2.66	N	45[c]	5.41
Potassium chloride	April 10, 1976	2.70[b]	1.56	K	60[c]	2.60
Dolomitic lime	April 10, 1976	0.75[b]	0.43	Dolomitic lime	100	0.43
Organic fertilizer[d]	July 17, 1974	0.25[e]	0.31	Cow manure equivalent	50[f]	0.61

[a] Cruzeiro values corrected to January 1, 1975 using inflation of 35%/year.
[b] Prices of Brasil Norte Ltda., Altamira (where financed colonists purchase supplies).
[c] Coelho and Varlengia 1972:181.
[d] Rice bran and spoiled beans.
[e] Average of Cr$0.30/kg for rice bran and Cr$0.20/kg for spoiled beans paid by a colonist of Japanese heritage who was using 2,222 kg/ha of each of these on black pepper.
[f] Estimated (for purposes of cost) from the fact that the dose for the rice bran and spoiled bean mixture (note e) was double the recommended manure dose for pepper.

TISOL) with an initial pH of 3.9 (Struchtemeyer et al. 1971:22). The pH changes for the regression were 28 days after application, and only lime doses up to 8,000 kg/ha were used (Fearnside 1978:549–50). A maximum limit of 9.0 is assumed, above which liming cannot raise pH.

Changes in soil aluminum from liming are calculated using the same regression of aluminum on pH and clay content used for generating aluminum values for soil under virgin forest (equation A.1). The data on changes in aluminum are not included in the report of the Zona Bragantina liming trials, although a hand-sketched curve of the trend in aluminum (Struchtemeyer et al. 1971: graph 5) indicates that a relation with pH holds which is very similar to the relation found between aluminum and pH in virgin soil in the present study. Values for aluminum are, of course, restricted from taking on negative values in the simulation.

Changes in soil phosphorus from fertilization are calculated using the regression given by:

$$Y = 0.0568 \, A - 6.41 \qquad \text{Equation A.33}$$

where:

Y = change in phosphorus (ppm)
A = P_2O_5 applied (kg/ha)
($p < 0.0001$, $r = 0.99$, $N = 14$).

The data used for the regression in equation A.33 come from experiments in the cerrado zone of Brazil (North Carolina 1974: 89 and 101). A maximum limit of 100 ppm is assumed, above which soil phosphorus cannot be raised through fertilization.

The fact that a patch has been fertilized is stored for use in temporarily adjusting the soil carbon levels as a result of manuring (an effect which dissipates within one year) in the black pepper yield subroutine.

Only cacao and pepper are fertilized in the simulation. Pasture is not fertilized in the simulation, nor is it fertilized by actual colonists on the Transamazon Highway.

Soil Erosion

Soil erosion, operationally defined for the purposes of modeling soil fertility changes as the lowering of the soil surface, can be predicted based on slope, weather information, and soil composition. This measure of erosion actually includes soil compaction as well, but this does not detract from the measure's usefulness in predicting fertility changes, since fertility changes can be shown empirically to be related to this measure of erosion and other factors through regression analysis.

Erosion was measured in the area using a series of 47 plots of stakes (Fearnside 1980d). Each plot was an array of 15 lengths of plastic pipe driven into the ground with a notch cut in the stake at ground level. The fall or rise in ground level could then be measured at a later time using a ruler.

Erosion rates depend on land use. The land use on the day with the maximum rainfall in a 24-hour period was used. Average plot erosion in sites either bare (less than 60 days uncultivated) or in annual crop fields of different types (rice, maize, manioc, and *Phaseolus* beans) were all similar. The regression used in the simulation for patches in these conditions is given by:

$$Y = 0.164\,A + 1.88 \times 10^{-3}\,B + 1.43 \qquad \text{Equation A.34}$$

where:

Y = plot erosion (mm/year)
A = plot slope (%)
B = rain while bare or in annual crops (mm)
($p < 0.0001$, $r = 0.89$, SE = 2.13, N = 17 plot means).

For other uses the erosion at individual stakes was used. Stake slope (slopes over distances of 30 cm) is related to plot slope (slopes over distances of about 20 m) by:

$$Y = 0.462\, A + 313 \qquad \text{Equation A.35}$$

where:

Y = stake slope (%)
A = plot slope (%)
($p<0.001$, $r = 0.89$, SE = 4.94, N = 705 stakes, 47 plots, df = 45).

Erosion under black pepper is given by:

$$Y = 0.712\, A + 6.05 \qquad \text{Equation A.36}$$

where:

Y = stake erosion (mm/year)
A = stake slope (%)
($p<0.001$, $r = 0.55$, SE = 6.10, N = 39 stakes).

Stake erosion measures during the observation period, which was about one year in all cases, are shown in table A.19 for land uses for which regressions could not be developed.

Table A.19. Soil Level Decreases Under Other Uses

Land Use	Mean (mm/yr)	SD	N (stakes)
Weeds (2–8 months fallow)	8.1[a]	5.4	56
Second growth (> 8 months fallow)	6.9[a]	8.7	68
Pasture	6.7[a]	11.8	105
Young cacao	10.0	8.3	40
Virgin forest	7.5[b]	5.1	75

SOURCE: Fearnside 1980d.
[a] Not significantly different. Variances differ significantly ($p < 0.0001$, F = 25, df = 3,144710).
[b] Believed to be high.

Crop Yields

MAIZE

The maize yield regression and associated multipliers are derived from an initial data set of 224 maize fields. This was first culled to remove fields with: 1) areas less than one hectare; 2) questionable data due to noted contradictions or vagueness in colonist responses; 3) cases where soil sample did not come from the location of the field but from a similar nearby location with identical history; and 4) cases with incomplete data for any of the regression variables: soil pH, planting density, interplanted rice density, and interplanted manioc density. Excluded categories of valid data were cases with rat damage of intensities 3 or 4; cases with poor germination noted; and cases with disease noted.

In cases where maize planting density was not known from direct field measurement, the density was estimated using a regression of density on the weight of seeds planted per hectare forced through the origin. This regression is given by:

$$Y = 414.88 \, A \qquad \text{Equation A.37}$$

where:

Y = maize planting density (plants/ha)
A = seeds sown (kg/ha)
($p<0.0001$, $r=0.65$, SE $=56.40$, N $=75$).

For the regression predicting maize yields from soils and interplanting information, maize yields are given directly in kilograms per thousand plants. The pH values are adjusted to 6.0 in accord with the linear response and plateau model (Waugh et al. 1975).

The regression equation for maize yield prediction is given by:

$$Y = 125.46 \, A - 2.92 \times 10^{-2} \, B - 2.22 \times 10^{-2} \, C - 8.16 \times 10^{-4} \, D - 330.00 \qquad \text{Equation A.38}$$

where:

Y = maize yield (kg/1,000 plants)
A = pH (adjusted to 6.0)
B = maize planting density (plants/ha)
C = interplanted manioc density (plants/ha)
D = interplanted rice density (plants/ha)
($p<0.05$, $r=0.65$, SE $=151$, N $=28$).

PHASEOLUS BEANS

Planting density in plants per hectare is generated from the regression in equation A.39, which is forced through the origin:

$$Y = 1602.8 \, A \qquad \text{Equation A.39}$$

where:

Y = "bean" (*Phaseolus* or *Vigna*) density (plants/ha)
A = weight of seeds sown (kg/ha)
($p<0.0001$, $r=0.82$, SE $=55966$, N $=28$).

Table A.20. *Phaseolus* Bean Yield Regression Variables Summary

Item	Mean	SD	N
kg seeds planted/ha	29.18	26.15	112
Planting density (plants/ha)	52,666	49,036	114
Interplanted maize:			
frequency	11.7%		120
maize density (plants/ha)	1,698.4	3,688.2	13
Frequency of planting in previously planted soil when virgin soil available	9.1%		55

The above regression was performed lumping 20 *Phaseolus* and 8 *Vigna* fields. When similar regressions were performed on the smaller data set for *Phaseolus* alone, a coefficient of 1595.5 was obtained ($p<0.0001$, $r=0.82$, $SE=257.15$, $N=20$).

The data for the regression of *Phaseolus* yields and associated adjustments come from an original data set of 120 *Phaseolus* fields. This was culled to remove fields with areas less than one hectare; questionable data (noted contradictions or vagueness in colonist responses) for production, disease, or area; and incomplete data for yield, density, interplanted maize density, disease, or soil pH. Excluded categories of valid data were fields with disease of any intensity, and fields with poor germination reported. For use in the regression, pH values were adjusted to 5.7. The critical value of pH 5.7 comes from Coelho and Verlengia (1972:133). *Phaseolus* yield is given by:

Table A.21. *Phaseolus* Bean Yield Regression Excluded Condition Summary

Item	Mean	SD	N
Poor germination:			
frequency	5.8%		120
effect on yield[a]	1.002		1
Disease:			
Frequency			
overall	67%		115
virgin soil	78%		50
previously planted	100%		5
Effect on yield	0.3496	0.5728	32

[a] An *a priori* decision was made to exclude poor germination from all crop yield regressions.

$$Y = 267.64 - 69.765\,A + 13.777\,B - 1.50 \times 10^{-3}\,C \qquad \text{Equation A.40}$$

where:

> Y = *Phaseolus* yield (kg/kg seed sown)
> A = planting density ($\log_{10}$ plants /ha)
> B = pH (adjusted to 5.7)
> C = interplanted maize density (plants/ha)
> ($p < 0.05$, $r = 0.79$, SE $= 29.80$, N $= 13$).

VIGNA COWPEAS

The yield regression and multiplier values are derived from an initial data set consisting of 34 *Vigna* fields. This was culled by eliminating: fields less than 0.5 ha in area; questionable data for production, density, or area; cases where sample did not come from the *Vigna* field but from a nearby field with identical history; fields with production estimated prior to sacking (no actual cases); and fields with incomplete data for production, area, disease, or soil pH. Excluded categories of valid data were: fields with disease of any intensity; fields with germination problems reported (no actual cases); and fields with rabbit attack of intensity 3 or 4 (heavy or very heavy).

For the purposes of modeling, simulated pH values were adjusted to 6.0. No actual cases in the data set used for the yield regression had pH values this high, so no adjustment was made in the data analysis. The regression for predicting *Vigna* yields is given by:

Table A.22. *Vigna* Cowpea Yield Regression Variables and Excluded Conditions Summary

Item	Mean	SD	N
	Variables		
Kg seeds planted/ha	8.10	7.82	30
	Excluded conditions		
Disease			
frequency	14.3%		28
effect on yield	0.0876		1
Rabbits (intensity 3 or 4)			
frequency	17.65		34
effect on yield	0.677		1
Poor germination			
frequency	0%		
effect on yield	?		

$$Y = 20.81 \, A - 84.40 \qquad \text{Equation A.41}$$

where:

$Y = $ Vigna yield (kg/kg seed sown)
$A = $ pH (adjusted to 6.0)
($p = 0.054$, $r = 0.87$, SE $= 13.895$, N $= 5$).

BITTER MANIOC

Bitter manioc growth period is generated from the mean growth period of 472 days and standard deviation of 193 days (N $= 64$). For growth periods of less than one year the growth period multiplier is generated from the mean proportion of the 1–2 year old manioc yield: 31.8 (SD $= 0.40$, N $= 7$). The corresponding mean multiplier for fields over two years old is 0.40 (SD $= 3.7$, N $= 3$).

The data for the bitter manioc yield regression and multiplier calculations come from an initial data set of 64 fields. This was culled to eliminate fields less than 0.5 ha in area, and fields with incomplete data for growth period, yield, or area. Excluded categories of valid data for the regression were the fields with less than one year and over two years of growth. No adjustments or transformations were performed on the data used in the regression, but the pH values were adjusted to 5.0 in the simulation since this was the highest pH value present in the culled data set. This is also a reasonable value for the critical pH above which further increases will not improve manioc yields. Soil pH values below 5.0 have been found to give reduced yields (Almeida and Canéchio Filho 1972:160).

The regression for predicting bitter manioc yields is given by:

$$Y = 4124.4 \, A - 17369 \qquad \text{Equation A.42}$$

where:

$Y = $ bitter manioc yield (kg flour/ha/12 months growth)
$A = $ pH (adjusted to 5.0)
($p < 0.05$, $r = 0.93$, SE $= 414.22$, N $= 5$).

SWEET MANIOC

Sweet manioc growth period is generated ($\bar{x} = 471$ days, SD $= 103$, N $= 11$). Then pH is adjusted to 5.0, this critical value being the same as that used in the case of bitter manioc. The data for the sweet manioc yield prediction regression and growth period adjustment are drawn from an initial data set of 13 sweet manioc fields. The data set was culled by eliminating fields with areas less than 1.0 ha. A larger minimum area was required for sweet manioc than for bitter manioc due to bias in small fields from the practice of harvesting sweet manioc little by little for eating boiled or feeding to pigs. Unfortunately, not only are larger areas needed for reliable data, but since plantings of sweet manioc are generally smaller than is the case for bitter manioc, many fields were eliminated. Also eliminated were any fields with incomplete data for production, growth period, area, or pH. Excluded categories of valid data were fields with growth periods less than one year and fields with growth periods over two years (of which there were no actual cases). The regression obtained is given by:

$$Y = 587.53 \ A - 1559.2 \qquad \text{Equation A.43}$$

where:

Y = sweet manioc yield (kg flour/ha/12 months growth)
A = pH (adjusted to 5.0)
($p = 0.26$, $r = 0.92$, SE = 81.50, N = 3).

Several reasons exist for using the above regression despite the lack of statistical significance and the small sample size. The regression parallels the results for bitter manioc, except that sweet manioc yields are lower. The regression also confirms field impressions.

In the 15 percent of cases where the growth period is less than one year, the growth period multiplier ($\bar{X} = 3.77$, SD = 3.74, N = 2) adjusts yields to reflect the faster growth rate of the tubers during the first year of growth.

Pasture and Cattle

The relation used for predicting grass production of *Panicum maximum* from soil phosphorus for use in modeling was derived from experiments done with another grass species, *Brachiaria decumbens*, in Belém (Serrão et al. 1971). The soil at the site of the Belém experiments was also a yellow latosol (ULTISOL) low in phosphorus (2 ppm). In deriving the value of the base yield used (303.0 kg dry weight/ha/year), a correction was made for the difference in production between *Brachiaria decumbens* and *Panicum maximum* using a factor of 1.12 (Viégas and Kass 1974:33).

Yields were calculated as proportions of the base yield of pasture grass.

The base yield represents what would be expected for *Panicum maximum* in Altamira on soil with a total phosphorus level of 2 ppm.

The weed competition correction used comes from data available from *Brachiaria* experiments done in Belém, from which the proportion of the total dry weight made up of inedible weeds in successive years as unfertilized pasture could be calculated (Simão Neto et al. 1973:9). The proportions for the fourth and fifth years were assumed (optimistically) to be equal to that of the third year.

The simulated grass dry weight production is then converted into beef production. This is done by first converting the dry weight of grass into total digestible nutrients, using a factor of 0.54 (based on Vicente-Chandler 1975:424, see Fearnside 1979a:222). Production of total digestible nutrients/ha/year can be converted to live weight gain of cattle/ha/year using a factor of 0.14, based on a conversion factor used by Vicente-Chandler (1975:424) for the reverse calculation from live weight gain to pasture production, attributed to "Pasture Research Committee 1943." Feeding capacities can be calculated from the average slaughter weight in Amazonia of 330 kg and mean age at slaughter of four years (do Nascimento and de Moura Carvalho 1973:III-B-32).

In order to reach this weight in four years, the cattle must grow at a rate of 82.5 kg/head/year. A three-year feeding capacity calculated at the 2 ppm soil phosphorus level would be derived from the base yield multiplied by the year effects for weed competition, averaged over the three years and converted to total digestible nutrients, yielding a three-year average of 187 kg total digestible nutrients/ha/year for *Panicum maximum* on the Transamazon Highway. Converted to live weight gain, the beef production would be 26.2 kg weight gain/ha/year averaged over three years. This production, divided by the 82.5 kg/head/year which the cattle must gain to reach slaughter weight on schedule, gives a three-year feeding capacity of 0.32 head/ha. If "animal units" of 350 kg each are used in place of the 330 kg "head" weight used here, the corresponding feeding capacities are lowered by 6.1 percent.

Dry weight of pasture grass produced per hectare is calculated from the base pasture yield, the year effect, and the coefficient and constant from the pasture yield regression on phosphorus (Fearnside 1979a). The regression on phosphorus is given by:

$$Y = 4.84 \, A \qquad \text{Equation A.44}$$

where:

$Y =$ (pasture yield with phosphorus/pasture yield without phosphorus) $- 1.0$

A = phosphorus (ppm) − 2.0
Note: phosphorus range is 0<A<8 ppm
($p<0.01$, r = 0.82, SE = 1.27, N = 8).

The regression in equation A.44 is based on data from a fertilization experiment on *Brachiaria decumbens* in Belém (Serrão et al. 1971). Yields are calculated for the fertilized plots at the time of each of the eight clippings of the experimental plots as proportions of the corresponding yields in the plot with the same treatment except for the addition of phosphorus. The critical value for phosphorus response was estimated at 10 ppm using the strongest phosphorus response from *Brachiaria* fertilizer trials done in Brasília (North Carolina 1974:101), and phosphorus levels above this were taken to have the same effect as the critical value in accord with the linear response-plateau model (Waugh et al. 1975). All phosphorus values used in the regression were converted to values in excess of the unfertilized phosphorus level of 2.0 ppm, and yields were expressed in proportions of the control yield so that the regression could be forced through the origin. The relations of pasture yield with phosphorus and weed effects are combined to give:

Y = A (4.84 B − 8.68) C   for phosphorus <10 ppm
Y = 39.72 A C   for phosphorus ≥10 ppm
Equation A.45

where:

Y = pasture yield (kg dry weight/ha/year)
A = base yield (expected first year yield in kg dry weight/ha for *Panicum maximum* at 2.0 ppm phosphorus; here equal to 303 kg/ha/year)
B = soil phosphorus (ppm)
C = year factor (proportional decrease from first year yield due to invasion of weeds). Values: first year: 1.00; second year: 0.63; third year: 0.49; fourth year: 0.49; fifth year: 0.49.

Cattle live weight gain is calculated from pasture dry weight yield by multiplying this by the weight gain per kg total digestible nutrients (0.14) and the total digestible nutrients as a proportion of the dry matter weight (0.54).

Black Pepper

Black pepper yields are predicted from a multiple regression on soil pH, carbon, and phosphorus (Fearnside 1980a), based on data from EMBRAPA fertilizer trials in Belém (de Albuquerque and Condurú 1971:110). Soil nutrient levels in the plots with various fertilizer treatments had to be estimated from analyses done on other experiments being conducted at the same location. Control plot nutrients are taken from Serrão et al. (1971:10): pH = 4.7, aluminum ions = 1.2 meq/100g, carbon = 0.94 percent, nitrogen = 0.07 percent, exchangeable phosphorus = 4 ppm. Soil analysis methods were the same as those used for the samples taken on the Transamazon Highway for the present carrying capacity study. Critical values used in adjusting high nutient values in the data analysis and simulation were: pH = 5.5 (de Albuquerque and Condurú 1971:98), phosphorus = 10 ppm (the level for "high" fertility used by EMBRAPA for pepper fertilizer recommendations: Brazil, IPEAN 1966), carbon = 2.0 percent. The carbon value was chosen above the 1.2 percent critical carbon for most crops (Catani and Jacintho 1974:33–34) due to strong yield responses obtained with manuring (de Albuquerque and Condurú 1971:110) despite a high 0.94 percent carbon level for the soil type (Serrão et al. 1971:10). Yields in the fertilizer trials were converted to proportions of the maximum yield for the year. The regression is given by:

$$Y = 0.292\ A + 0.382\ B - 0.0552\ C - 2.119 \quad \text{Equation A.46}$$

where:

Y = black pepper yield (proportion of maximum yield for year)
A = pH (adjusted to 5.5)
B = carbon (% dry weight, adjusted to 2.0)
C = phosphorus (ppm, adjusted to 10.0)
($p < 0.0001$, $r = 0.86$, SE = 0.187, N = 24).

Variability is introduced in the simulation through the standard error of the estimate. Adjustments of regression-predicted yields to reflect expectations in Altamira are made by multiplying the predicted yield in its form as a proportion of the maximum yield by an official figure for the fertilized yields expected, in this case 5500 kg dry seeds/ha (de Albuquerque et al. 1973:26). Year effects used to adjust for immature plants were: 1 year = 0.00; 2 years = 0.40; 3 years = 0.80; 4 or more years = 1.00 (de Albuquerque et al. 1973:26). Plants die in the simulation at an age of 12.5 years, the midpoint of Morais' estimate of 10–15 years for life expectancy (1974b:7.5).

The effect of *Fusarium* disease is incorporated through a multiplier expressing the proportion of the healthy yield expected. A value of 0.5 is used, based on the assumption that plants are killed at a constant rate when a patch is attacked, and that individual plants die instantaneously. The probability per year of *Fusarium* establishment in an area given that it is not already established is 0.206 (mean time to entry = 3 years, N = 2 areas); the probability of attacking a patch given that the disease is established in the area is estimated to be 0.293 (average years to occur = 2), and the probability that a patch is killed given that it is diseased is estimated to be 0.206 (average years to occur = 3) (Fearnside 1980a).

Product Allocation

NONAGRICULTURAL INCOME

Hunting is a supplementary income source for those colonists whose cultural backgrounds lead them to engage in this activity. If a simulated colonist is a hunter, game catch must be calculated for the lot, the colonist's labor supply must be adjusted for the time spent in hunting, and game sold to other colonists must be accounted for with appropriate additions to the hunter's cash supply for the proceeds from the sales.

Hunting is only carried out during the period for which game populations are considered to survive at a level which will contribute a significant amount to the colonists' supply of meat. The value used for the end of hunting is 15 years, based on Smith's discovery that Agrovila Coco Chato near Marabá, settled fifteen years previously, had only 761 kg/year of game harvested (1976b:171–73). An alternative value would be twenty-one years, the point at which yield would be zero at the present rate of decline.

The total effort spent in hunting for the year is calculated from:

$$Y = 84.00 - 2.00 \text{ A} \qquad \text{Equation A.47}$$

where:

Y = the total hunting effort for the year (man-days/hunter/year)
A = the number of years since the beginning of hunting (since the beginning of the simulation).

Table A.23. Outside Labor Type Frequencies

Colonist Type	Labor Type					
	Daily wage	Enterprise	Government or professional	Working women or children	Probability No Labor	Sample Size
Entrepreneur	0.33	0.33	0.17	0.17	0.00	6
Independent farmer	0.11	0.29	0.00	0.11	0.49	28
Artisan farmer	0.38	0.04	0.17	0.04	0.37	24
Laborer farmer	0.39	0.09	0.06	0.09	0.37	65

NOTE: Differences between labor types: $p < 0.0001$, $\chi^2 = 52.9$, DF = 18, N = 121.

Table A.24. Outside Labor: Days Spent and Earnings by Labor Type

Labor Type	Total Labor			Male Labor			Earnings		
	Mean[a]	SD	N	Mean[a]	SD	N	Mean	SD	N[c]
Daily wage	30.6	41.5	93	30.6	41.5	93	16.25[b]	(0)	
Enterprise	17.5	12.0	2	17.5	12.0	2	0.0920[d]	0.7754	5
Government/professional	34.0	22.9	3	34.0	22.0	3	8420.5[e]	6929.8	7
Working women and children	217.4	205.2	7	108.9	229.4	7	6346.2[e]	6929.8	7

[a] Labor in man-days/lot/year spent in each labor type given that a colonist engages in that labor type. Male labor is for males 18 years or older.
[b] Daily wage earnings in cruzeiros of January 1, 1975/ adult male man-day.
[c] Average of 12 months at the going wage rates in 1975 (also used as "cost of labor" input): 9 months at Cr$15/man-day and 3 months (felling) at Cr$20/man-day (without food included). Standard deviation is assumed to be zero, since there is not much variation in wages paid between lots at any one time of year.
[d] Earning return on investments in cruzeiros of January 1, 1975 / cruzeiro of January 1, 1975 invested, considering investment as both capital and capital goods lumped.
[e] Earnings in cruzeiros of 1, January 1975/year.

Table A.25. Cash Sent and Received from Outside Area

Event	Probability per Year	N (colonist years)	Value, Given > 0 (Cr75$/yr)		
			Mean	SD	N
Cash received from outside	0.053	263	17,788	20,344	9
Cash sent to outside	0.011	263	492	121	3

NOTE: US$1 = Cr75$7.4.

Equation A.47 is derived from data on hunting efforts collected by Smith in two agrovilas three and fifteen years after settlement respectively (1976b:171–73, 195).

The game yield per man-day is calculated from:

$$Y = 14.13 - 0.68 A \qquad \text{Equation A.48}$$

where:

Y = game yield in kg/man-day
A = hunting effort in man-days.

Equation A.48 is derived from the observations of Smith, that in a three-year-old agrovila the effort was 373 man-days/agrovila/year or 78 days/hunter/year at 3.5 hunters/agrovila, while in an agrovila on a site which had been settled for fifteen years the effort was 190 man-days/agrovila/year or 54 days/hunter/year, with yields in the newer settlement at 12.1 kg/man-day vs. 4.0 kg/man-day in the older settlement (1976b:171–73).

Next, the subsistence need for game for the month is calculated. The subsistence need per month in dressed weight is the subsistence need for meat of any kind (kg/person/year) divided by 12 months and multiplied by family size. The subsistence need for meat is discussed in the section on product allocation in chapter 4.

The hunting effort expended in each month of the year is determined, with appropriate adjustments being made in family (total) labor and male labor supplies. All hunting labor is considered to be male labor. The hunting effort in each month is calculated from the yearly total hunting effort for the lot and the proportion of the hunting effort which falls into each month. These proportions have been calculated from the data given by Smith (1976b:357) for hunting effort in Agrovila Nova Fronteira, 90 km west of Altamira. The proportions are: January: 0.026; February: 0.082; March: 0.084;

April: 0.148; May 0.104; June: 0.108; July: 0.124; August: 0.136; September: 0.016; October: 0.014; November: 0.072; December: 0.086.

The bag obtained in each month is calculated as the effort expended in the month times the yield per man-day of hunting effort. This live weight figure for game catch is adjusted to obtain a dressed weight figure using a game wastage factor of 0.40 (Smith 1976b:173). The dressed weight captured in each month is compared with the subsistence need for game per month. If the amount captured in the month is less than or equal to the subsistence need, then all of the game is eaten and none is sold. If the monthly bag is sufficient to meet subsistence need, then any excess game is added to the community pool of game available for purchase by other colonists.

SPOILAGE OF PRODUCTS

Table A.26. Spoilage of Stored Products

Item	Kept for Consumption or Sale			Kept for Use as Seeds		
	Mean Proportion spoiled	SD	N	Mean Proportion spoiled	SD	N
Rice	0.180	0.250	39	0.10	0.17	5
Maize	0.275	0.286	21	0	(0)	1
Phaseolus	0.051	0.092	13	0	(0)	1
Vigna	0	0	3	0	(0)	1
Bitter manioc	0	0	(assumed)	n.a.		
Sweet manioc	0	0	(assumed)	n.a.		
Cacao	0	0	(assumed)	n.a.		
Pepper	0	0	(assumed)	n.a.		

TRANSPORT TO MARKETS

Given that transportation is available, transporting products to market (CIBRAZEM) costs a mean of Cr75$0.12/kg (SD = 0.08, N = 11 colonists). If colonists sell their product to the Banco do Brasil, they must make trips to the Bank in the town of Altamira (50 km away) to collect payment. The mean cost of each trip is Cr75$37.54 (SD = 33.32, N = 6). The optimistic assumption is made that only two trips are necessary to collect payment.

Table A.27. Transportation Availability

Zone	Probability Lot in Zone[a]	Probability No Transport to Market		N (km)
		First 3 years	Later years	
1. Roadside	0.26	0.00	0.00	15.00
2. First 10 km of side roads	0.40	0.16	0.00	16.75
3. End of side roads	0.34	1.00	0.23	11.50

[a] Proportion of 232 lots assigned to colonists in study area (excluding the 4 *agrovilas*).

Buying and Selling of Products

Not all products are sold to the Banco do Brasil. The frequencies of sale to the bank from 1975 were used in the simulation. The probability of selling rice to the bank was 0.433 (N = 12 colonists), maize was 0.000 (N = 3), and *Phaseolus* was 0.000 (N = 1). The other crops are not purchased by the bank in Altamira, even if the colonist were to wish to sell these. When crops are sold to the bank, in addition to the cost of trips to receive payment, the bank automatically deducts payments of principal and interest due on previous loans.

Table A.27a. Prices of Products

	Buying (Cr75$/kg)[a]			Selling (Cr75$/kg)[a]		
Product	Mean	SD	N	Mean	SD	N
Rice (in husks)	2.44	0.84	12	1.25	0.24	10
Maize (grain)	0.85	0.13	3	0.74	0.24	2
Phaseolus beans	10.62	4.15	12	3.64	1.17	3
Vigna cowpeas	8.85	(3.46)	—[b]	3.02	0.98	—[b]
Bitter manioc (flour)	2.90	0.60	13	1.87[c]	(0.00)	1
Sweet manioc (flour)	2.90	0.60	—[d]	1.87[c]	(0.00)	—[d]
Cacao (dry seeds)	—	—	—	6.00	(0.00)	official
Black pepper	—	—	—	7.52	(0.00)	1
Game (dressed weight)	9.23	2.18	5	9.23	2.18	5
Canned meat	14.93	4.53	8	—	—	—

[a] US$1 = Cr75$7.4.
[b] Calculated from price relation of *Vigna* to *Phaseolus* prices on February 15, 1975, when *Vigna* sold for 83.3 percent of the price of small-grained *Phaseolus* varieties (such as "preto") or 62.5 percent of the price of large-grained varieties (such as "jalo").
[c] Before a deduction of 30 percent for flour-making equipment (*casa de farinha*) use.
[d] Assumed equal to bitter manioc.

Subsistence Needs for Agricultural Products

The subsistence quantities of each crop have been based on the midpoint of the quantities of the crop consumed by buyers and sellers of the crop interviewed during fieldwork. For rice (with hulls), buyers consumed a mean of 105.5 kg/person/year ($SD = 57.0$, $N = 16$), and sellers consumed a mean of 175.7 kg/person/year ($SD = 82.3$, $N = 3$). The subsistence need for rice used was therefore 140.6 kg/person/year. In the case of maize there were no sellers in the data set. The mean for buyers was 195.5 kg/person/year ($SD = 264.8$, $N = 4$), which was used as the subsistence need for maize (including maize used for chicken feed). This corresponds to 14.55 kg of chicken eaten per person per year, or a standing crop of 23.84 birds/capita or 143.07 birds for a family of six persons. This higher than the observed average flock of 51.3 chickens/family ($SD = 35.3$, $N = 8$). This is partly compensated for by the omission of rice as feed in the simulation. For "beans" (both *Phaseolus* and *Vigna*) the average consumption among buyers was 43.9 kg/person/year ($SD = 19.6$, $N = 12$), and among sellers was 34.6 kg/person/year ($N = 1$). The midpoint of 39.3 kg/person/year was used as the subsistence need for "beans." For manioc (both sweet and bitter), buyers ate a mean of 43.4 kg/person/year of flour ($SD = 31.8$, $N = 13$), while sellers ate 80.0 kg/person/year ($N = 1$). The midpoint of 61.7 kg/person/year of flour was used as the subsistence need for manioc.

Subsistence Need for Cash

The subsistence cash need includes money for purchases to meet any deficit in seed storage if stocks set aside from the lot's own production are insufficient. A subsistence cash need also exists for items that cannot be produced on the lot, such as clothing, medicine, and transportation. The subsistence cash need for these items totals Cr75$1615.30/person/year, being composed of Cr75$293.53 for clothing ($SD = 297.21$, $N = 8$ colonists), Cr75$129.91 for kerosene, matches, soap, etc. ($SD = 129.91$, $N = 11$), Cr75$145.63 for medicines ($SD = 214.47$, $N = 10$), Cr75$1029.21 for foods not grown in the lot (including meat) ($SD = 486.30$, $N = 6$), Cr75$709.02 for foods grown on the lot ($SD = 246.54$, $N = 7$), Cr75$133.34 for transportation of people (other than for arranging for financing, payment of installments to the bank, and collecting payments from the bank for products sold to the government) ($SD = 107.59$, $N = 15$), and Cr75$34.61 for other consumption ($SD = 69.53$, $N = 7$), less Cr75$151.91 for canned meat and game, to which cash is allocated separately.

Table A.28. Prices and Nutrition of Principal Goods Bought with Cash

		Units/capita/yr			Price (Cr75$/ unit)	Cost (Cr75$/ capita/ yr)[b]	Nutrition/unit[a]			Nutrition/capita/day		
Item	Unit	Mean	SD	N			Calories	Total protein	Animal protein	Calories	Total protein (g)	Animal protein (g)
Sugar	kg	42.12	23.93	14	3.98	167.64	3,695	0	0	426.10	0	0
Salt	kg	8.95	1.70	14	1.70	15.22	0	0	0	0	0	0
Powdered milk	kg	4.11	2.77	13	30.02	123.38	5,080	250	250	57.16	2.81	2.81
Kerosene	liters	12.39	7.78	16	3.41	42.25	0	0	0	0	0	0
Coffee	kg	5.15	3.07	14	18.17	93.58	0	0	0	0	0	0
Oil	liters	11.62	4.54	13	14.20	165.00	10,131	0	0	322.30	0	0
Soap	kg	10.60	6.06	13	6.82	72.29	0	0	0	0	0	0
Macaroni	kg	2.26	4.65	8	9.09	20.54	3,552	0	0	21.98	0	0
TOTALS										827.54	2.81	2.81

[a] Brazil, ACAR-PARÁ n.d. [a] (1974).
[b] Average family size in full sample = 6.33 (SD = 1.40, N = 15).

Game, aside from that eaten directly by hunters' families, can be purchased from the community pool of game available for sale, which comes from what the hunters in the community kill beyond the amount eaten by their families. Meat requirements are satisfied first by converting maize to chickens and from game already present, then by purchasing game if available, and last from the purchase of canned meat.

Colonists' diets are made up of the calories, total protein, and animal protein gained from the quantities of staples eaten, from game and canned meat, from the contributions of other products normally bought with cash, and from the tuber crops such as the part of sweet manioc production consumed directly as tubers, as well as sweet potatoes *(batata doce* : *Ipomea batatas)*, yams *(inhame* or *cará* : *Dioscorea* spp.), and to a small extent tannia *(taioba*: *Xanthosoma* spp.). The consumption of tubers, considered a "free" supplement to the diet, averages 480.19 kg/capita/year (N=4 lots), or 135.39 kg/capita/year after discounting waste using a factor of 71.94 percent for sweet manioc tubers (de Carvalho nd [1974]). This tuber consumption level contributes 176.01 thousand calories/capita/year (SD=70.33, N=4 lots), and 1761 g/capita/year of total protein (SD=703, N=4 lots). The animal protein contribution is zero.

The contributions from products bought for cash, given the observed colonist spending patterns, total Cr75$5.34/1000 calories, Cr75$1573.85/kg total protein, and Cr75$1573.85/kg animal protein. Table A.28 gives the prices and nutritional contributions of the principal goods purchased with cash, outside of canned meat, game, products which are also produced in the colonists' lots, and "other consumption" included in subsistence cash need. Product prices for items produced in the lot were given earlier in table A.27.

Debt Payments

Colonists who have not sold their produce to the Banco do Brasil often do not pay their bank debts voluntarily. The observed probability was 0.00 (N=3), but a probability of 0.25 is assumed for the purposes of simulation since the formal sample is inadequate and informal impressions lead me to expect a somewhat higher value. "Voluntary" payment is only made given that some money is available after subsistence needs have been satisfied. These low repayment probabilities represent a substantial subsidy to the colonists, both in the simulation and on the Transamazon Highway. Less lenient bank and INCRA policies from 1977 onwards with regard to colonists who do not pay their debts have raised these probabilities on the Transamazon Highway.

The probability of paying a private loan, such as goods received on credit from colonist shops, loans from friends, etc., is even lower. The observed probability was 0.00 (N = 2), but a value of 0.10 was assumed as a more realistic estimate of how often these loans are repaid. There is great variability among colonists on the Transamazon Highway, both in how likely borrowers are to repay loans and in how extreme lenders are in inducing them to pay. The overall probability of repayment is low, in any case.

Seed Requirements

Table A.29. Seed Requirements

Crop	Seed Requirement (kg/ha)		
	Mean	SD	N
Upland rice	29.6	13.4	263 fields (any combination)
Maize	12.5	9.4	66 fields (maize alone)
Phaseolus beans	29.2	26.1	112 fields (any combination)
Vigna cowpeas	8.1	7.8	30 fields (any combination)

Investment

Colonists' labor classification determines whether they invest their money in their lots or in some other enterprise. Colonists whose *labor* pattern *(not colonist type)* is as an entrepreneur or professional (or either of these in combination with other labor activities) invest only a portion of their available funds in their lots. The proportion used in the simulation is 0.74 (SD = 0.38, N = 6). Otherwise all money slated for investment goes to lot development.

Table A.30. Proportions of Free Capital Invested After Satisfying Subsistence Needs

Colonist Type	Proportion of Free Capital		
	Mean	SD	N
Entrepreneur	0.36	0.17	3
Independent farmer	0.46	0.40	3
Artisan farmer	0.22	0.25	6
Laborer farmer	0.13	0.13	2

NOTE: Free capital is cash not spent on subsistence crops, canned meat, or game.

Table A.31. Proportions of Investment Capital Within Each Category Spent on Purchase of Capital Goods

Colonist Type	Use of Capital Good (category)		
	Lot	Enterprise	Consumption
Entrepreneur	0.33	0.22	0.03
	(2)	(3)	(2)
Independent farmer	0.05	0.05	0.05
	(1)	(3)	(1)
Artisan farmer	0.07	(0.00)	(0.00)
	(3)	(0)	(0)
Laborer farmer	0.00	0.00	0.00
	(1)	(1)	(1)

NOTE: Numbers in parenthesis are sample sizes.

Capital goods depreciate at a rate estimated at 20 percent per year. This is based on a weighted average of the depreciation rates of power saws and threshing machines in the study area, as shown in table A.33. This figure agrees with the depreciation rate used by INCRA for carts (Brazil, INCRA 1972a:194).

Population Processes

INITIAL POPULATION

Table A.32. Initial Population Characteristics (at arrival)

Item	Mean	SD	N
Lot owner's age (years)	37.03	10.78	123[a]
Probability wife present	0.898	—	108
Wife's age if present (years)	31.87	10.67	112[a]
Probability dependents present other than wife, given colonist is married	0.912	—	91
Probability other dependents present given colonist is single	0.182	—	11[a]
Number of dependents, other than the wife, given at least one present	4.14	2.27	103[a]
Probability sex of dependent, other than wife, is male	0.53	—	456[a]

[a] Includes some newcomers in the sample; no differences apparent.

Table A.33. Depreciation of Capital Goods

Item	Expected Life (years)	Depreciation/ Year	Sample Sizes Machines	Sample Sizes Years	Proportion of Colonists	Sample Size (colonists)	Value (Cr75$ each)[a]	Proportion of Total Capital Goods	Weighted Depreciation
Power saws	4	0.25	6	2	0.098	61	7,588	0.58	0.14
Threshers	8	0.13	4	2	0.026	152	21,000	0.42	0.06
Weighted depreciation rate									0.20

[a] US$1 = Cr75$7.4.

Table A.34. Return on Capital Goods Relative to Manual Labor

Item	Unit	Maintenance (Cr75$/unit)	SD	N	Fuel (Cr75$/unit)	SD	N	Cost of Labor with Machine (Cr75$/unit)[b]	SD	N	Total Costs (Cr75$/unit)[a]	Value of Labor If Done Manually (Cr75$/unit)	SD	N	Average Relative Return	Weighted Relative Return
Power saws	ha felled	90.98	19.19	3	107.84[c]	—		46.60	25.00	7	245.42	186.80	73.00	12	0.76	0.44
Threshers	50 kg rice sack				0.47		1	1.13[d]	0.35	2	1.60[e]	4.65	4.20	5	2.91	1.23
Weighted relative return rate																1.67

[a] Exclusive of depreciation.
[b] At Cr75$20/ man-day for felling, Cr75$15/man-day for other tasks.
[c] Calculated from Brazil, INCRA (1972a:193) (16 l gasoline + 4 l oil/ha).
[d] Calculated from Moran (1975;127), counting child labor as 50 percent of adult equivalent.
[e] Not including threshing machine maintenance costs (unavailable).

Table A.35. Demographic Information

Demographic Age Class	Dependent Age Dist. (other than wife)[a]	Individual Immigration Probabilities		Individual Emigration Probabilities		Probability of Death Under Adequate Nutrition	
		Male	Female	Male	Female	Male	Female
1	0.077	0.000	0.000	0.0000	0.0000	0.0156	0.0118
2	0.074	0.000	0.000	0.0000	0.0000	0.0008	0.0005
3	0.064	0.000	0.000	0.0000	0.0000	0.0008	0.0005
4	0.065	0.000	0.111	0.0000	0.0000	0.0008	0.0005
5	0.052	0.000	0.000	0.0000	0.0000	0.0008	0.0005
6	0.052	0.000	0.111	0.0000	0.0000	0.0006	0.0004
7	0.055	0.000	0.000	0.0000	0.0000	0.0006	0.0004
8	0.036	0.000	0.111	0.0000	0.0000	0.0006	0.0004
9	0.039	0.000	0.111	0.0000	0.0000	0.0006	0.0004
10	0.039	0.000	0.111	0.0000	0.0000	0.0006	0.0004
11	0.061	0.000	0.000	0.0000	0.0000	0.0004	0.0003
12	0.055	0.000	0.000	0.0000	0.0000	0.0004	0.0003
13	0.039	0.000	0.000	0.0000	0.0000	0.0004	0.0003
14	0.047	0.000	0.111	0.0000	0.0000	0.0004	0.0003
15	0.017	0.000	0.000	0.0000	0.0000	0.0004	0.0003
16	0.025	0.000	0.000	0.0000	0.0323	0.0010	0.0004
17	0.036	0.000	0.000	0.0000	0.1379	0.0010	0.0004
18	0.028	0.067	0.111	0.0303	0.0345	0.0010	0.0004
19	0.011	0.000	0.000	0.1034	0.0000	0.0010	0.0004
20	0.022	0.067	0.000	0.0625	0.0000	0.0010	0.0004
21	0.017	0.067	0.000	0.0968	0.0000	0.0011	0.0004
22	0.017	0.200	0.000	0.0968	0.0714	0.0011	0.0004
23	0.008	0.067	0.000	0.0690	0.0000	0.0011	0.0004
24	0.014	0.000	0.000	0.0385	0.0000	0.0011	0.0004
25	0.003	0.133	0.000	0.0000	0.0000	0.0011	0.0004
26	0.008	0.000	0.000	0.0000	0.0000	0.0008	0.0005
27	0.000	0.067	0.000	0.0000	0.0000	0.0008	0.0005
28	0.000	0.067	0.111	0.0000	0.0000	0.0008	0.0005
29	0.006	0.000	0.000	0.0000	0.0000	0.0008	0.0005
30	0.003	0.000	0.000	0.0000	0.0000	0.0008	0.0005
31	0.000	0.000	0.000	0.0435	0.0000	0.0010	0.0007
32	0.000	0.067	0.000	0.0000	0.0435	0.0010	0.0007
33	0.003	0.000	0.000	0.0000	0.0000	0.0010	0.0007
34	0.000	0.000	0.111	0.0000	0.0000	0.0010	0.0007
35	0.000	0.000	0.000	0.0000	0.0000	0.0010	0.0007
36	0.000	0.000	0.000	0.0000	0.0000	0.0016	0.0011
37	0.000	0.000	0.000	0.0000	0.0000	0.0016	0.0011
38	0.000	0.000	0.000	0.0000	0.0000	0.0016	0.0011
39	0.000	0.000	0.000	0.0000	0.0000	0.0016	0.0011
40	0.000	0.000	0.000	0.0000	0.0000	0.0016	0.0011
41	0.000	0.000	0.000	0.0000	0.0000	0.0026	0.0022
42	0.000	0.000	0.000	0.0000	0.0000	0.0026	0.0022
43	0.000	0.000	0.000	0.0000	0.0000	0.0026	0.0022

MODEL PARAMETERS

Demographic Age Class	Dependent Age Dist. (other than wife)[a]	Individual Immigration Probabilities		Individual Emigration Probabilities		Probability of Death Under Adequate Nutrition	
		Male	Female	Male	Female	Male	Female
44	0.000	0.000	0.000	0.0000	0.0000	0.0026	0.0022
45	0.000	0.000	0.000	0.0000	0.0000	0.0026	0.0022
46	0.000	0.000	0.000	0.0000	0.0000	0.0040	0.0033
47	0.000	0.000	0.000	0.0000	0.0000	0.0040	0.0033
48	0.000	0.000	0.000	0.0714	0.0000	0.0040	0.0033
49	0.000	0.067	0.000	0.0000	0.0000	0.0040	0.0033
50	0.000	0.000	0.000	0.0909	0.0000	0.0040	0.0033
51	0.000	0.000	0.000	0.0000	0.0000	0.0058	0.0041
52	0.000	0.000	0.000	0.0000	0.0000	0.0058	0.0041
53	0.000	0.000	0.000	0.0000	0.0000	0.0058	0.0041
54	0.000	0.000	0.000	0.0000	0.0000	0.0058	0.0041
55	0.000	0.000	0.000	0.0000	0.0000	0.0058	0.0041
56	0.000	0.000	0.000	0.0000	0.0000	0.0107	0.0066
57	0.000	0.000	0.000	0.0000	0.0000	0.0107	0.0066
58	0.000	0.000	0.000	0.0000	0.0000	0.0107	0.0066
59	0.000	0.067	0.000	0.0000	0.0000	0.0107	0.0066
60	0.000	0.000	0.000	0.0000	0.0000	0.0107	0.0066
61	0.000	0.000	0.000	0.0000	0.0000	0.0186	0.0107
62	0.003	0.000	0.000	0.0000	0.0000	0.0186	0.0107
63	0.000	0.000	0.000	0.0000	0.0000	0.0186	0.0107
64	0.000	0.000	0.111	0.0000	0.0000	0.0186	0.0107
65	0.000	0.000	0.000	0.0000	0.0000	0.0186	0.0107
66	0.006	0.000	0.000	0.0000	0.0000	0.0295	0.0183
67	0.000	0.000	0.000	0.0000	0.0000	0.0295	0.0183
68	0.000	0.000	0.000	0.0000	0.0000	0.0295	0.0183
69	0.003	0.000	0.000	0.0000	0.0000	0.0295	0.0183
70	0.000	0.000	0.000	0.0000	0.0000	0.0295	0.0183
71	0.000	0.000	0.000	0.0000	0.0000	0.0478	0.0346
72	0.000	0.000	0.000	0.0000	0.0000	0.0478	0.0346
73	0.000	0.000	0.000	0.0000	0.0000	0.0478	0.0346
74	0.000	0.067	0.000	0.0000	0.0000	0.0478	0.0346
75	0.000	0.000	0.000	0.0000	0.0000	0.0478	0.0346
76	0.000	0.000	0.000	0.0000	0.0000	0.0683	0.0564
77	0.000	0.000	0.000	0.0000	0.0000	0.0683	0.0564
78	0.000	0.000	0.000	0.0000	0.0000	0.0683	0.0564
79	0.000	0.000	0.000	0.0000	0.0000	0.0683	0.0564
80	0.000	0.000	0.000	0.0000	0.0000	0.0683	0.0564
81	0.000	0.000	0.000	0.0000	0.0000	0.1175	0.1000
82	0.000	0.000	0.000	0.0000	0.0000	0.1175	0.1000
83	0.000	0.000	0.000	0.0000	0.0000	0.1175	0.1000
84	0.003	0.000	0.000	0.0000	0.0000	0.1175	0.1000
85	0.000	0.000	0.000	0.0000	0.0000	0.1175	0.1000
86	0.003	0.000	0.000	0.0000	0.0000	0.1981	0.1875

[a] N = 362 dependents, lumped for original and newcomer colonists.

Table A.36. Initial Capital and Capital Goods

Colonist Type	Liquid Capital			Capital Goods for:								
				Lot			Enterprises			Consumption		
	$\bar{X}$	SD	N	$\bar{X}$	SD	N	$\bar{X}$	SD	N	$\bar{X}$	SD	N
Original colonists												
1. Entrepreneur	0	0	1	0	0	3	0	0	3	0	0	1
2. Independent	5,265	11,431	14	460	4,210	23	1,166	4,210	23	285	1,067	14
3. Artisan	1,768	3,186	23	0	0	22	0	0	22	97	456	22
4. Laborer	2,360	6,958	55	0	0	61	0	0	61	0	0	55
Newcomer colonists												
1. Entrepreneur	7,849	11,099	2	0	0	2	0	0	2	0	0	2
2. Independent	26,296	30,257	9	1,174	3,523	9	2,979	6,533	9	587	1,762	9
3. Artisan	0	0	1	0	0	1	0	0	1	0	0	1
4. Laborer	1,383	2,461	6	0	0	6	0	0	6	0	0	6

NOTE: All values in cruzeiros of January 1, 1975 (Cr75$7.4 = US$1.00).

HUNTER STATUS

Not all colonists hunt for game in the forest; many who have come from areas of Brazil outside of Amazonia are either fearful of the forest and unskilled in its ways, or are distainful of hunting as a low-status pursuit. The handful of colonists who hunt regularly have supplied an important part of the animal protein consumed by the community during the first years of colonization on the Transamazon Highway.

Each colonist's status as a hunter or a nonhunter is assigned in the simulation based on the probability of 0.07 that a colonist is a hunter—a value calculated from Smith's statement based on surveys in three agrovilas that there are 2–5 hunters/agrovila (1976:171–73). The probability was calculated using 3.5 hunters/agrovila (the midpoint of the range of Smith's observations), and assuming the usual 50 houses as the agrovilas' mean size.

FAMILY LABOR AND HEALTH

Table A.37. Labor Equivalents in Agricultural Work

Age Group (years)	Man	Woman
7–8	0.20	0.15
9–13	0.25	0.20
14–17	0.50	0.40
≥ 18	1.00	0.75

SOURCE: Brazil, INCRA 1972a:202.
NOTE: Proportion of adult male contribution.

Table A.38. Disease Probabilities by Age and Sex

Age Group (years)	Malaria Male	Malaria Female	Trauma Male	Trauma Female	Other Male	Other Female
0–4	0.0227	0.0140	0.0027	0.0016	0.0405	0.0260
5–9	0.0185	0.0098	0.0021	0.0009	0.0058	0.0021
10–14	0.0103	0.0038	0.0033	0.0005	0.0018	0.0016
15–19	0.0214	0.0090	0.0062	0.0011	0.0021	0.0005
20–24	0.0198	0.0116	0.0084	0.0011	0.0038	0.0011
25–29	0.0138	0.0191	0.0053	0.0006	0.0024	0.0032
30–34	0.0175	0.0176	0.0095	0.0021	0.0009	0.0021
35–39	0.0130	0.0196	0.0065	0.0039	0.0059	0.0065
40–44	0.0144	0.0025	0.0058	0.0012	0.0023	0.0037
45–49	0.0072	0.0165	0.0046	0.0033	0.0033	0.0033
50–54	0.0248	0.0248	0.0038	0.0000	0.0019	0.0000
55–59	0.0248	0.0062	0.0248	0.0062	0.0331	0.0062
≥ 60	0.0165	0.0248	0.0414	0.0000	0.0083	0.0124

NOTE: Table shows probability that a given individual contracts disease at some time during a given year. Probabilities calculated from numbers of hospital admittances in SESP hospital in Altamira by sex and age class for 1973 given by Smith (1976b:217, 235, 239, 247). Proportions of individuals in the entire area by sex and age class were estimated from the survey of 101 families done by Moran (1975) as modified by Smith (1976b:216) to include Smith's estimates of individuals in each category in the (mostly migrant labor) population not under INCRA auspices (Smith 1976b:216). The estimate of the total population of the area used in the calculations is 34,000 (Smith 1976b:212). (Note: Smith also gives a figure of 24,000 [p. 205]).

Table A.39. Monthly Disease Probabilities

Disease	1	2	3	4	5	6	7	8	9	10	11	12
Malaria[a]	0.061	0.083	0.126	0.067	0.056	0.051	0.067	0.104	0.155	0.092	0.081	0.057
Trauma[b]	0.051	0.080	0.048	0.070	0.090	0.049	0.079	0.103	0.081	0.103	0.141	0.105
Other[c]	0.083	0.084	0.083	0.083	0.084	0.083	0.084	0.083	0.083	0.084	0.083	0.083

NOTE: Table shows probability that hospital internment occurs during a given month given that the individual contracts the disease at some time during the year.
[a] Malaria probabilities calculated from SUCAM data for Altamira area on numbers of positive slides found for malaria in blood samples from patients with suspected malaria referred to this service. Calculated from data in Smith (1976b:212).
[b] Trauma (accident) probabilities calculated from numbers of hospital admittances in SESP hospital, Altamira, for 1971, 1972, and 1973. Data from complete census of hospital records done by N. Smith was taken from Moran (1976:121).
[c] Other diseases: assumed equal to 1/12 for each month due to lack of data. Rounding differences assigned randomly.

Table A.40. Work Days Lost to Illness

Disease	Days Lost[a]			Other Values	
	Mean	SD	N	Smith(1976b)	Moran(1975)
Malaria	33.3	23.1	3	14	8
Trauma	86.7	79.0	6	19[b]	
Other	45.0	30.0	3	14.5[c]	

[a] Actual work days lost, not man-day equivalents.
[b] Based on estimate of 3 times average hospital stay, which was 6.2 days.
[c] Based on 3 times average hospital stays of 4.8 days for gastroentristis and 4.9 days for respiratory diseases, and equal probabilities of attack for each (0.5 percent of population per year for each disease).

FERTILITY AND MORTALITY

The probability of death from each nutrient deficiency is calculated from

$$P = (A-1)\left(1 - \frac{B}{C}\right) \quad \text{Equation A.49}$$

where:

P = the probability of death per year from deficiency of the nutrient (calories or protein)
A = the mortality factor for the nutrient (calories or protein)
B = the average amount of the nutrient eaten per capita per day (calories or protein)
C = the nutrient requirement per capita per day (calories or protein).

Following the procedure of Weisman (1974), only one of the two nutrients is considered limiting. Whichever gives the higher probability of nutrient

Table A.41. Age-Specific Fertility for Rural Brazilian Population

Age Group	Total No. Women	Total Live Births in Previous Year	Probability of Live Birth for Age Group
15–19	2,188,350	132,029	0.060
20–24	1,772,913	414,291	0.234
25–29	1,334,934	362,629	0.272
30–34	1,087,643	274,233	0.252
35–39	987,395	198,659	0.201
40–44	837,890	91,988	0.110

SOURCE: 1970 census figures presented in da Mata et al. (1973:175).

Table A.42. Calorie and Total Protein Requirements and Effects on Mortality

Age (years completed)	Calorie Requirement[a] (per capita/day)	Calorie Mortality Factor[b]	Total Protein Requirement[c] (g/capita/day)	Total Protein Mortality Factor[d]
0	820	2.6	11.9	3.35
1	1,180	2.35	12.3	2.35
2	1,355	2.35	12.3	2.20
3	1,540	2.35	12.3	2.05
4	1,695	2.1	15.4	1.95
5	1,830	2.1	15.4	1.85
6	1,955	2.1	15.4	1.80
7	2,075	1.8	19.2	1.70
8	2,185	1.8	19.2	1.60
9	2,295	1.8	19.2	1.55
10	2,400	1.55	23.1	1.50
11	2,475	1.55	23.1	1.45
12	2,550	1.55	23.1	1.40
13	2,625	1.25	26.2	1.33
14	2,700	1.25	26.2	1.23
15	2,750	1.25	26.2	1.15
16	2,735	1.25	26.2	1.12
17	2,720	1.05	26.2	1.08
18	2,785	1.05	26.2	1.00
19	2,610	1.05	26.2	1.00
20–39	2,600	1.00	25.4	1.00
40–49	2,470	1.00	25.4	1.00
50–59	2,430	1.00	25.4	1.00
60–69	2,080	1.00	25.4	1.00
70+	1,820	1.00	25.4	1.00

[a] From Weisman (1974:B485), who derived this from UN, WHO (1973).
[b] A factor expressing the effect of calorie deprivation on excess mortality in populations by age class. Derived from Weisman (1974:B493).
[c] Weisman (1974:B488) derived from UN, WHO (1973). In egg protein equivalent.
[d] A factor expressing the effect of total protein deficiency on excess mortality by age class. Derived from Weisman (1974:B495).

deficiency-related death is used. Note that the nutrient consumption levels are yearly averages over all family members.

Colonist Marriage

In modeling marriages, only those of lot owners are considered explicitly. Children marrying and leaving the lot are considered as individual emigrants, while spouses of children or others joining them in the parents' lot are considered as individual immigrants. The yearly probability of a single colonist (lot owner) marrying is 0.069 (N = 29 single colonist-years). The mean age of the brides marrying lot owners was 21 years (SD = 9, N = 2). This figure is probably higher than it would be if an adequate sample were available.

IMMIGRATION AND EMIGRATION

Immigration and emigration are divided into two types: individual and family level moves. These have very different implications for the development of the area. When individuals arrive to join already established families, as by marriage (of persons other than the lot owner, which is treated separately), the result is an increase in the labor force (and product demand) without changes in the exploitation patterns of the lot. Colonists' sons and daughters leaving for cities, to establish new homesteads or to marry, likewise changes family size without affecting basic strategies. When whole families are replaced, however, the behavioral differences of newcomer colonists are felt.

The probability per year of a lot receiving an individual immigrant is 0.049 (N = 491 lot-years, 24 individual immigrants). The probability that an individual immigrant received by a lot is male is 0.625 (N = 24). Given the individual immigrant's sex, the probability that the person belongs to each age class can be determined from the age- and sex-specific immigration probabilities given with the demographic information in table A.35.

Table A.43. Probability per Year of Family Emigration

Colonist Type	Single Colonists			Married Colonists		
	Probability	N[a]	Number of Colonists Leaving	Probability	N[a]	Number of Colonists Leaving
1. Entrepreneurs	0.143[b]	0	0	0.063	16	1
2. Independent farmers	0.200	5	1	0.012	81	1
3. Artisan farmers	0.143	7	1	0.057	88	5
4. Laborer farmers	0.105	19	2	0.022	229	5

[a] Colonist years.
[b] Assumed equal to artisan farmers.

Notes

1. Development Rush in the Amazon Rainforest

1. The Legal Amazon shown in figure 1.1 is after a slight modification in its southern boundary on January 1, 1979. The new boundary is an irregular line slightly south of the 16th parallel, which had served as the boundary since designation of the area in 1953 (Decree Law 1806). Figures for the old area have ranged from 4,990,570 km^2 (de Almeida 1977) to 5,033,072 km^2 (Rebelo 1973:43). The new area is 5,005,425 km^2 if water surfaces are included and 4,975,607 km^2 if they are excluded (Brazil, IBGE 1982:28; see table 1.2). Another value reported for terrestrial area is 4,975,527 km^2 (Tardin et al. 1980).

2. Grainger 1980, Lanly and Gillis 1980, Myers 1979, 1980a, b; NRCCRPTB 1980; Sommer 1976; U.S. DS 1978; UNEP 1980.

3. Tardin et al. (1978, 1979) surveyed 552,000 km^2 in the general area of the Belém-Brasília Highway, and found that 41,000 km^2 (7.4 percent) had been felled by the time the images were taken in 1977. The study includes a large area of *cerrado* (Central Brazilian scrubland) vegetation type, even including areas at the latitude of Brasília in the sample.

4. Speech reproduced in Tamer 1970:249–53.

5. People of mixed Amerindian and Caucasian descent or, more generally, any poor Portuguese-speaking inhabitants of the Amazonian interior.

6. Early official references to a goal of one million families were quickly reduced by INCRA to the 100,000 figure (Kohlhepp 1980:60).

7. Described by Sanders 1973; Wesche 1974; Goodland and Irwin 1975a; Kleinpenning 1975; Smith 1976a, b, 1981, 1982, and Moran 1976, 1981.

8. This practice differs from the custom among some Amerindian groups in lowland South America. In Venezuela, Harris found maize planted on better-cleared ground than manioc (1971:492). On the Transamazon Highway manioc is usually planted after the ground has already undergone a *coivara* for a previous rice crop, and it is maize that is relegated to the tangle of branches when a burn is poor and little time remains for *coivara*.

9. This law (Law No. 2,597, art. 8 of 1955) is sometimes circumvented by registering several contiguous parcels in the names of different family members.

2. The Tropical Rainforest As an Ecosystem

1. In 1966 Falesi estimated there were 25,000 km^2 of *terra roxa* known (1967:156). By 1970 this figure had fallen by more than half to 10,600 km^2 (Falesi and Rodrigues 1970:3), by 1973 it was 10,300 km^2 (Falesi 1974a:209), and by 1974 it was 10,000 km^2 (Falesi 1974b:2.8).

2. The analogy of carbon dioxide with a greenhouse is somewhat misleading as the latter has most of its effect as a barrier to convection rather than to long-wave radiation.

3. Some investigators have reported findings implying a substantially lower effect on global temperatures from a given increase in atmospheric CO_2. Using meteorological observations accompanying a volcanic eruption as a "natural experiment" rather than computer simulations used in other studies, Newell and Dopplick (1979) estimate that temperature increases resulting from a doubling of atmospheric CO_2 would be no more that 0.25°C in tropical latitudes. Another experimental study (Idso 1980a, b), using short-term observations over a continental land mass, calculates a value for global mean warming of up to 0.26°C. A number of investigators regard these estimates as low (P. J. Crutzen, pers. comm. 1980; Leovy 1980; Schneider et al. 1980; G. M. Woodwell, pers. comm. 1980). The difference in the results is apparently due to omission of the positive feedback between ocean temperature and tropospheric warming caused by radiation absorption and re-radiation by water vapor (Kerr 1982; Ramanathan 1981; NAS 1982a).

3. Population Growth and Carrying Capacity

1. Ammerman (1975) has published an excellent discussion of archaeologists' approach to the problem of carrying capacity estimation.

2. Brush (1975) has independently shown the identity of the Allan and Carneiro formulas, but reached erroneous conclusions regarding the formulas of Gourou and Conklin. One error apparently arose from failure to see the erratum slip in the front of the journal containing the original publication of Conklin's (1959) work correcting a misprint in the formula (T was printed in place of L in the denominator of the critical population density formula.) Brush also mistakenly interpreted the A in the formula as acreage rather than area in hectares but left the conversion constant of 100 in the equation, which had served to convert hectares to square kilometers in Conklin's (1959) formulation. In the case of Gourou's formula, Brush claimed that the rotation period should not be simply the cultivated period plus the fallow period, but should be this quantity divided by the cultivated period. However, Gourou's other published discussion of the formula (1971:188) confirms his original intention, and the original Gourou formula can be readily shown to be identical to the other shifting cultivation formulas.

3. Some anthropologists have argued that various population-limiting behaviors evolved to hold population sizes below carrying capacity for the benefit of groups (e.g., Meggers 1971). Basing carrying capacity determinations on individual consumptions should encourage framing analyses of this kind in terms of the inclusive fitness of individuals (Durham 1976; Lewontin 1970; Williams 1966).

4. Some studies combine a group of factors implicitly through terms such as "the amount of food that [an average individual] ordinarily derives from cultivated plants per year" (Carneiro 1960; see Box). Animal protein is usually not included in these combined measures.

5. See, for example, Allan 1949, 1965; Brookfield and Brown 1963: 110–14;

Fearnside 1972; Rappaport 1968; Waddell 1972:170; and UNESCO/UNFPA 1977:386–89.

6. This type of distortion is sometimes referred to as a "Procrustean bed," named after the giant of Greek mythology who seized travelers and tied them to an iron bedstead, then cut off their legs or stretched them until they fit the bed.

7. Such confusion has clouded public debate of the WORLD3 "limits-to-growth" models of global resource, pollution, and population trends (D.H. Meadows et al. 1972, 1973; D.L. Meadows et al. 1973).

8. Statisticians use the terms "variable" and "parameter" somewhat differently than do modelers. Statisticians speak of "independent" variables, factors (regardless of origin with respect to system boundaries) influencing a "dependent" variable whose value the investigator would like to describe or predict. "Parameters" to statisticians refer to characteristics of the universe under study, as distinguished from statistics approximating these values based on estimates from sampling.

4. Modeling the Agroecosystem of Transamazon Highway Colonists

1. Dense forest of the low plateau subregion of Pará/Maranhão/Amapá; dense submontane forest with dissected topography of Carajás; dense plateau forest of the high plateau subregion of the Xingú/Tapajós; and open mixed forest of the leveled surface subregion of the mid Xingú/Irirí (Brazil, RADAMBRASIL 1974, vol. 5, mapa fitoecológico, folha SA.22).

2. Falesi (1972a) classifies much of the same area as red-yellow podzolic (ULTISOL).

3. In this study, the term *virgin forest* refers to forest not cleared by colonists, since Amerindians had been present in the area for centuries prior to the colonization program.

4. Graner et al. (1972) give a value of 1.7 percent carbon. Some values are lower still, such as the 1.5 percent organic matter (corresponding to 0.87 percent carbon) considered adequate by the Brazilian Soil Testing Service for Minas Gerais (cited North Carolina 1974:149).

5. A Stochastic Model for Human Carrying Capacity on the Transamazon Highway

1. Note that the values and dates shown on the program output do not represent predictions for particular years. The time scale shown on output graphs is intended only as a guide to the lengths of the runs' time horizons.

2. The 25 g/person/day UN-FAO standard used for animal protein is quite high, 10 g/person/day being considered adequate by many authorities, or even less depending on a variety of factors (McArthur 1977). The higher figure is consistent with the standards used by local government organs (Brazil, ACAR-PARÁ 1974). The high priority Brazilians place on meat (included in the product allocation sector of KPROG2) insures that animal protein intake is more than adequate from a nutritional standpoint if sufficient resources are available.

3. As in the discussion of inputs in chapter 4, all cruzeiro values used in the program are corrected for inflation to January 1, 1975. At that time the minimum monthly wage in Pará was Cr$326.40 and the exchange rate for the U.S. dollar was 7.4. These units are designated "Cr$75."

4. Areas shown are planted areas for all crops with the exception of bitter and sweet manioc, which refer only to the area harvested. Rice areas are total areas for both rice alone and interplanted. Maize area is total of maize alone and interplanted maize with correction for density to give equivalent of maize planted alone (using a correction factor of 0.647 for interplanted maize, calculated from Fearnside 1978:580 as a weighted average).

5. All runs in figure 5.12 have the dynamic population sector with the exception of the run with population restricted to laborers only at a density of 60 persons/km^2, which had the population sector frozen. The runs for which failure probabilities are shown for individual criteria in figure 5.11 are those at densities of 64.0 and 71.3 persons/km^2.

Appendix: Summary of KPROG2 Model Parameters and Equations

1. For comparison with Moran's Cr75$753–1329 range, Bunker calculates a range equivalent to Cr75$138–3506, based on 1977 costs for transport and other needs in three different colonization projects, including Altamira (1980b:587). Bunker's estimate includes food and lodging costs (Cr75$30–721) not included by Moran, while Moran includes trips for confirmation of loan approval (Cr75$57–95) not included by Bunker (Bunker 1980b:587; Moran 1976:86).

2. Bunker reports range for three colonization projects of 3–19 days (1980:587), as compared with Moran's range of 17–30 days (1976:86), including 3–5 days for confirmation of approval, not included in the Bunker (1980b:597) range.

Glossary of Technical Terms, Foreign Words, and Abbreviations

ACAR-PARÁ. (Associação de Crédito e Assistência Rural do Estado do Pará). Association for Credit and Rural Assistance of the State of Pará.

Agregado. An "attached one"; a person who is not related to the colonist living on a lot.

Agricultural Year. July 1 (beginning of *broca*) through June 30 (end of rice harvest).

AGRISIM. A program that uses the agricultural production sector of KPROG2 to simulate yields of individual crops in use on the Transamazon Highway.

Agroecosystem. Ecological systems that include the crops and other organisms cultivated or managed to supply the food and other needs of humans. May include one or more crops and the humans who are supported by those crops, depending on the investigator's purpose in studying the system. As used for the KPROG2 models, the Transamazon Highway agroecosystem includes several crops and crop combinations plus the human population.

Agropolis. Planned agricultural towns on the Transamazon Highway, which include a commercial zone and offices of government functionaries.

Agrovila. Planned agricultural villages on the Transamazon Highway with approximately fifty houses built by INCRA.

Amerindian. Indigenous peoples of the New World ("Indians").

Amazonia. The Amazon region. Can refer to either the Brazilian Legal Amazon, the Amazon Basin regardless of national boundaries, or the Hylean portion of Brazil. *See also* Legal Amazon.

Banco do Brasil. Bank of Brazil.

BASA (Banco da Amazônia, Sociedade Anônima). Bank of Amazonia, Anonymous Society.

Behavior mode. Broad patterns of system behavior such as population stability, collapse, etc.—in contrast to the prediction of precise values for different variables at specific times.

Belém-Brasília Highway. The highway (BR-010) running north to south connecting Belém, at the mouth of the Amazon River, with the national capital at Brasília.

Broca. Underclearing, or cutting the vines, small poles, and understory plants in a piece of virgin rainforest in preparation for felling.

Caboclo. 1) Amazonian Brazilian of mixed Amerindian and Caucasian descent; 2) any poor Portuguese-speaking inhabitant of the Amazonian interior, regardless of racial background. *Caboclos* are born and raised in the Amazonian interior.

Campina. Open white sand scrub forest in the Rio Negro basin.

Campinarana. Closed white sand forest in the Rio Negro basin.

Capina. Weeding a crop after it has been planted. Not to be confused with *limpa*, or preparing an area for planting.

Capoeira. Second growth at least eight months uncultivated.

Carrying Capacity. 1) "The maximum number of persons that can be supported in perpetuity on an area, with a given technology and set of consumptive habits, without causing environmental degradation" (Allan 1949); 2) for estimation with stochastic models, the maximum population density that can be supported over a long time period without the probability of colonist failure, as defined by a combined measure of various consumption and environmental criteria, exceeding a specified maximum acceptable probability of colonist failure. Changes in technology with time can be included within the limits of the particular mode. *See also* Logistic carrying capacity.

Casa da farinha. A rough shelter with grater, griddle, press, bins, and other equipment for processing manioc into flour.

cc. Cubic centimeter(s).

CEC. Cation exchange capacity (sum of Ca^{++}, Mg^{++}, Na^{++}, K^+, H^+, and Al^{+++} in meq/100g).

CEPA (Comissão de Planejamento Agrícola). Agricultural Planning Commission, an organ under the state or territorial Secretatiat of Agriculture.

CEPLAC (Comissão Executiva do Plano da Lavoura Cacaueira). Executive Commission of the Cacao Growing Plan.

Cerrado. Savanna with low scrubby xerophytic trees. Characteristic of the Central Brazilian Plateau.

CIBRAZEM (Companhia Brasileira de Armazenamento). Brazilian Storage Company.

Club of Rome. An international group of private citizens (mainly industrialists) that sponsored the Forrester-Meadows and Mesarovic-Pestle world modeling groups.

Coevolved relationship. Relationship among different species evolved jointly through a long series of small changes. Many pollination and seed dispersal mechanisms of tropical rainforest trees involve highly specialized physical and behavioral adaptations that have coevolved between plants and animals.

Coivara. Piling up of unburned wood after the first burn in a field in preparation for a second burning.

Colonist failure. Failure of a colonist to meet one or more of the specified consumption or environmental quality criteria. Consumption criteria used are: family annual averages for per capita daily consumption of calories, total protein, animal protein, and cash standard of living in Cr75$/capita/month and in minimum wages/family/month. The environmental quality measure output is the proportion of the simulated colonist's lot cleared.

cm. Centimeter(s).

CPATU (Centro de Pesquisas Agropecuárias do Trópico Úmido). Center for Agriculture and Cattle Ranching Research of the Humid Tropics. (Branch of EMBRAPA in Belém). IPEAN prior to 1975.

Cr$. *See* Cruzeiro. Cr75$ = cruzeiros corrected for inflation to January 1, 1975.

Cruzeiro. Brazilian monetary unit (Cr$). Exchange rate in January 1975: US$1 = Cr$7.4. In April 1985: US$1 = Cr$4,500.0.

Cuiabá-Porto Velho Highway. The highway (BR;364) connecting Cuiabá, capital of Mato Grosso, with Porto Velho, capital of Rondônia.

Cuiabá-Santarém Highway. The highway running north to south connecting Santarém, at the confluence of the Tapajós and Amazon Rivers, and Cuiabá, the capital of Mato Grosso.

Custeio. Agricultural operations considered to be of transitory value, such as cleaning a field in preparation for planting *(limpa)*, planting, weeding, harvesting, and threshing. Cf. *Investimento*.

df. Degrees of freedom.

DENPASA (Dendê do Pará, Sociedade Anônima). Oil Palm of Pará, Anonymous Society.

Derruba. *See Derrubada*.

Derrubada. Felling of virgin forest.

Deterministic. Fixed; without random variation. A given set of initial conditions yields a specific outcome with a probability of one.

DNER (Departamento Nacional de Estradas de Rodagem). National Department of Highways.

Dynamic. Changing with time.

EMBRAPA (Empresa Brasileira de Pesquisa Agropecuária). Brazilian Enterprise for Agricultural and Cattle Ranching Research (includes IPEAN or CPATU).

EMBRATER (Empresa Brasileira de Assistência Técnica Rural). Brazilian Enterprise for Rural Technical Assistance; formerly ACAR-PARÁ in the state of Pará.

Exponential. Increasing (or decreasing) geometrically. Represented as a function of a number raised to a power. Exponential growth follows a J-shaped trajectory.

FAO. Food and Agriculture Organization of the United Nations.

Farinha. Manioc flour

Fecundity. Physiological capability to reproduce. *See also* Fertility.

Feeding capacity. The maximum density of stock which can be supported on a pasture for a given term.

Fertility. 1) Realized rate of reproduction; 2) in soil science, the measures of the suitability of soil chemical properties for agriculture.

Forrester-Meadows models. The world models developed at the Massachusetts Institute of Technology, under the sponsorship of the Club of Rome. These include the WORLD2 model (Forrester 1971), and the WORLD3 model (D. L. Meadows et al. 1973).

FORTRAN. "Formula Translation" computer language. First version was introduced by International Business Machines Corp. in 1957.

FUNAI (Fundação Nacional do Indio). National Indian Foundation.

g. Gram(s).

Gaucho. A person from the state of Rio Grande do Sul. Many are of German heritage.

Gleba. 1) In small colonization schemes, a group of about 70 lots, identified by a number; 2) in larger ranching schemes, a larger piece of land with a single owner, usually 500–3000 ha.

Grileiro. Land thief; a speculator who obtains land, often fraudulently arranging for expulsion of any previous occupants, and sells it at high profits.

Ha. Hectare(s) ($10,000m^2$, or $100m \times 100m$).

Hylea. The area floristically characteristic of the Amazon, excludes the parts of the Legal Amazon that are characteristic of the Central Brazilian Plateau or *cerrado* zone.

IAC (Instituto Agronômico de Campinas). Agronomic Institute of Campinas (São Paulo).

IBDF (Instituto Brasileiro de Desenvolvimento Florestal). Brazilian Instituto for Forestry Development.

IBGE (Instituto Brasileiro de Geografia e Estatística). Brazilian Institute of Geography and Statistics.

IDESP (Instituto do Desenvolvimento Econômico-Social do Pará). Instituto of Social and Economic Development of Pará.

Igapó. Swamp forest (found along black-water rivers).

IICA-TRÓPICOS (Instituto Interamericano de Cooperação Agrícola). Interamerican Institute for Agricultural Cooperation, formerly the Instituto Interamericano de Ciências Agrícolas (Interamerican Institute of Agricultural Sciences). IICA-TRÓPICOS, an organ of the Organization of American States, is the Belém branch of the institute (IICA) headquartered in Turrialba, Costa Rica.

INCRA (Instituto Nacional de Colonização e Reforma Agrária). National Institute for Colonization and Agrarian Reform.

Initialize. To assign a starting value to a variable.

INPA (Instituto Nacional de Pesquisas da Amazônia). National Institute for Research in the Amazon.

INPE (Instituto Nacional de Pesquisas Espaciais). National Space Research Institute.

Instantaneous. Applying to a given moment in time. An "instantaneous" rate does not imply that the rate can be sustained over a long period, or that the rate does not change with time.

IPEAN (Instituto de Pesquisas (e Experimentação) Agropecuárias do Norte). Institute for Agricultural and Cattle Ranching Research of the North. Became a part of EMBRAPA in 1974; renamed CPATU in 1976.

Iteration. Repetition of a series of calculations in a loop, the results of each pass through the chain of calculations or steps serving as the input for the succeeding repetition.

Jari (Companhia Florestal Monte Dourado S.A.). Monte Dourado Forestry Company, Anonymous Society, formerly Jari Florestal e Agropecuária, Ltda (Jari Forestry, Agriculture, and Cattle Ranching, Ltd.), the silviculture, irrigated rice, mining, and industrial estate originally owned by D. K. Ludwig on the Jari River.

Jeito. A bending of normal bureaucratic or legal procedures; a knack or trick.

K. 1) In population dynamics, the symbol for carrying capacity; 2) in soil chemistry, potassium.

km. Kilometer(s).

KPROG1. A preliminary program written for estimating human carrying capacity on the Transamazon Highway (Fearnside 1974).

KPROG2. The simulation program written for estimating human carrying capacity on the Transamazon Highway.

l. Liter(s).

LANDSAT. NASA satellites capable of producing false color images of all points on the earth's surface every 18 days; formerly called ERTS (Earth Resources Technology Satellite).

Legal Amazon. (Amazônia Legal). The federal territories of Amapá, and Roraima, the states of Pará, Amazonas, Rondônia, Acre, the new state of Mato Grosso, and the portions of Goiás north of 13°S. Latitude and Maranhão west of 44°W. Longitude. The new boundary between Mato Grosso and Mato Grosso do Sul is an irregular line slightly south of 16°S. Latitude, which formerly served as the southern limit of Legal Amazonia.

Licitação. Sale through closed tenders in which land is sold to the highest bidder with a bid above a government-established minimum.

Luso-Brazilian. Brazilian of Portuguese cultural tradition (regardless of racial background).

m. Meter(s).

Maximum acceptable probability of colonist failure. The highest proportion of colonists failing to meet one or more of the specified criteria that is acceptable for a planner.

meq/100 g. Milliequivalents per 100 grams of air-dried soil.

Mesarovic-Pestle models. The world models developed through the International Institute of Applied Systems Analysis in Laxenburg, Austria under the direction of Mesarovic and Pestle (1974a,b).

mg. Milligram(s).

mm. Millimeter(s).

Morador. A person or family not related to the colonist (lot owner) who lives on a lot; a "liver" or "dweller." *Moradores* are charged neither rent nor a portion of the crops, but the land they clear may be planted in pasture following one or two annual crops, thus increasing the use and resale values of the lot at no cost to the colonist. The *morador* may serve as a caretaker for the lot in the owner's absence.

N. 1) Population size; 2) sample size; 3) nitrogen (percent of dry weight of total N as determined by Kjeldhal method); 4) normal chemical concentrations.

Northern Perimeter Road. The still uncompleted highway paralleling the borders of Colombia, Venezuela, and the Guianas.

p. Probability that a relationship is due to random effects.

P. Total phosphorus.

PAD (Projeto de Assentamento Dirigido). Directed Settlement Project; a kind of planned colonization scheme with nominal government assistance. *See also* PIC.

Parameter. Constant external factors that influence a system.

Patches. Small pieces of land.

Phaseolus. Genus of true beans. Distinct from *Vigna* (cowpeas), which are also classed as "beans" in Portuguese.

PIC (Projeto Integrado de Colonização). Integrated Colonization Project, planned colonization project for small colonists (usually 100 ha lots) with a variety of services provided by the government.

Pilha. Long piles of harvested rice, with stalks, awaiting threshing.

PIN (Programa de Integração Nacional). National Integration Program.

POLAMAZONIA. (Programa de Pólos Agropecuários e Agrominerais da Amazônia). Program of Agricultural, Livestock, and Mineral Development Poles in Amazonia.

ppm. Parts per million.

PROBOR (Programa da Borracha). Rubber Program.

PRODEPEF (Programa para Desenvolvimento de Pesquisas Florestais). Program for Development of Forestry Research (a part of IBDF, discontinued in 1979).

r. 1) The innate rate of increase (instantaneous birth rate–instantaneous death rate); 2) regression coefficient.

r^2. Coefficient of determination.

RADAM (Radar na Amazônia). Radar in Amazonia Project. Renamed RADAMBRASIL when the project was expanded to cover all of Brazil.

Rainforest. Closed canopy broadleaf high biomass primary forest in the tropics. Technically many are classified as "tropical moist forest."

Roçagem. Cutting of second growth.

Rurópolis. Planned agricultural cities for the Transamazon Highway. Larger than an agrópolis, only one has been built—in PIC-Itaituba.

S.A. or S/A. (Sociedade Anônima). Anonymous Society (corporation).

Saturation density. *See* Logistic Carrying Capacity.

SD. Standard deviation.

SE. Standard error of the regression estimate.

SEMA (Secretaria Especial do Meio-Ambiente). Special Secretariat of the Environment.

SESP (Serviço Especial de Saúde Pública). Special Public Health Service.

Shifting cultivation formulas. Formulas for calculating carrying capacity for human populations supported on shifting cultivation using such information as farmed time, fallow time, and average yield and consumption.

Static. Not changing with time.

Stochastic. Probabilistic; including the effects of random variation in one or more factors.

SUDAM (Superintendência do Desenvolvimento da Amazônia). Superintendency of the Development of the Amazon.

SUDHEVEA (Superintendência do Desenvolvimento de Heveacultura). Superintendency for the Development of Rubber Culture.

SUFRAMA (Superintendência da Zona Franca de Manaus). Superintendency of the Manaus Free Trade Zone.

System. "A group of physical components connected or related in such a manner as to form and/or act as an entire unit" (Patten 1971:4).

Terra firme. Nonflooded uplands in the Amazon.

Terra roxa. "Purple earth." An ALFISOL, the best soil type on the Transamazon Highway with the exception of extremely limited patches of anthropogenic black soil *(terra preta do indio).*

Transamazon Highway. The highway (BR-320) that crosses the Brazilian Amazon from east to west from Recife and João Pessoa on the Atlantic Coast to Cruzeiro do Sul near the Peruvian border.

Travessão. Side roads perpendicular to the Transamazon Highway, usually extending about 20 km from the roadside; called *linhas* in Rondônia.

UNEP. United Nations Environmental Programme.

UNESCO. United Nations Educational and Scientific and Cultural Organization.

US$. United States dollars.

Várzea. Periodically inundated Amazonian floodplain along white-water rivers.

Vigna. Genus of cowpeas. In Portuguese these are classed together with *Phaseolus* as "beans" *(feijão)*.

Virgin forest. Primary rainforest not previously felled by colonists. May have been cleared previously by Amerindians, but no easily observable signs apparent in the superficial appearance of the vegetation (although soil profiles often contain charcoal).

$\overline{X}$. The mean.

Yr. year(s).

Zona Bragantina. The zone around the town of Bragança, near Belém in Pará. This 30,000 km^2 area, colonized at the end of the nineteenth century, supplied agricultural products to the rubber boom city of Belém.

Bibliography

Ackermann, F. L. 1966. A *depredação dos solos da Região Bragantina e na Amazônia*. Belém: Universidade Federal do Pará.
Ahn, P. M. 1979. The optimum length of planned fallows. In H. O. Mongi and P. A. Huxley, eds., *Soils Research in Agroforestry: Proceedings of an Expert Consultation Held at the International Council for Research in Agroforestry (ICRAF) in Nairobi, March 26–30, 1979*, pp. 15–39. Nairobi: ICRAF.
Allan, W. 1949. Studies in African land usage in Northern Rhodesia. *Rhodes Livingstone Papers*, no. 15.
Allan, W. 1965. *The African Husbandman*. New York: Barnes and Noble.
Almeida, T. de C. and V. Canéchio Filho. 1972. *Principais Culturas (2)*. Campinas, São Paulo: Instituto Campineiro de Ensino Agrícola.
Alvim, P. de T. 1973. Desafio agrícola da região Amazônica. *Ciência e Cultura* 24:437–43.
Alvim, P. de T. 1977a. The balance between conservation and utilization in the humid tropics with special reference to Amazonia, Brazil. In G. T. Prance and T. S. Elias, eds., *Extinction Is Forever*, pp. 347–52. New York: New York Botanical Garden.
Alvim, P. de T. 1977b. Possibilidades de expansão da fronteira agrícola nas regiões tropicais úmidas da América Latina. Paper presented at the VII Conferência Interamericana de Agricultura, September 5–7, 1977, at Tegucigalpa, Honduras. Instituto Interamericano de Ciências Agrícolas, Organização de Estados Americanos.
Alvim, P. de T. 1978a. Perspectives of agricultural production in the Amazon Region. *Interciencia* 3:243–51.
Alvim, P. de T. 1978b. Floresta amazônica: equilíbrio entre utilização e conservação. *Ciência e Cultura* 30:9–16.
Alvim, P. de T. 1978c. Floresta amazônica: equilíbrio entre utilização e conservação. *Silvicultura* 1978 (1): 30–35 and (2):54–59.
AAAS (American Association for the Advancement of Science). 1975. Food. *Science* 188:503–662.

Ammerman, A. J. 1975. Late pleistocene population dynamics: An alternate view. *Human Ecology* 3:219–33.

Andreae, B. 1974. Problems of increasing the productivity in tropical farming. *Applied Science and Development* 3:124–39.

Aschmann, H. 1959. *The Central Desert of Baja California: Demography and Ecology.* Berkeley, Calif.: Ibero-Americana, no. 42.

Ayres, J. M. 1977. A situação atual da área de ocorrência do cuxiú preto (*Chiropotes satanas satanas*—Hoff., 1807). Report to International Union for the Conservation of Nature (IUCN/SCC) Primate Specialist Group.

Ayres, J. M. 1978. *Reserva Ecológica de Aripuanã.* Relatório ao Instituto Nacional de Pesquisas da Amazônia (INPA).

Barcellos, J. M. 1974. Subsídios e diretrizes para um programa de pesquisa com bovinocultura na região norte. In *Reunião do Grupo Interdisciplinar de Trabalho sobre Diretrizes de Pesquisa Agrícola para a Amazônia (Trópico Úmido), Brasília,* May 6–10, 1974, 2:6.1–6.55. Empresa Brasileira de Pesquisa Agropecuária (EMBRAPA). Brasilia: EMBRAPA.

Barrett, S. W. 1980. Conservation in Amazonia. *Biological Conservation* 18:209–35.

Bartholomew, G. A., Jr., and J. B. Birdsell. 1953. Ecology and the protohominids. *American Anthropologist* 55:481–98.

Bayliss-Smith, T. P. 1974. Constraints on population growth: The case of the Polynesian outlier atolls in the precontact period. *Human Ecology* 2:259–95.

Bayliss-Smith, T. P. 1980. Population pressure, resources, and welfare: towards a more realistic measure of carrying capacity. In H. C. Brookfield, ed., *Population-Environment Relations in Tropical Islands: The Case of Eastern Fiji.* Man and the Biosphere (MAB) Technical Notes 13, pp. 61–93. Paris: UNESCO.

Beaver, S. E. 1975. *Demographic Transition Theory Reinterpreted.* Lexington, Mass.: Lexington Books.

Beinroth, F. H. 1975. Relationships between U.S. soil taxonomy, the Brazilian system, and FAO/UNESCO soil units. In E. Bornemsza and A. Alvarado, eds., *Soil Management in Tropical America: Proceedings of a Seminar held at CIAT, Cali, Colombia,* February 10–14, 1974, pp. 97–108. Raleigh: North Carolina State University, Soil Science Department.

Bennema, J. 1975. Soil resources of the tropics with special reference to the well-drained soils of the Brazilian Amazonian forest region. In *International Symposium on Ecophysiology of Tropical Crops, Manaus, May 25–30, 1975*, 1:1–47. Itabuna, Bahía: Comissão Executiva do Plano da Lavoura Cacaueira (CEPLAC).

Björkström, A. 1979. A model of CO_2 interaction between atmosphere, oceans, and land biota. In B. Bolin, E. T. Degens, S. Kempe, and P. Ketner, eds., *The Global Carbon Cycle*. Scientific Committee on Problems of the Environment (SCOPE) Report 13, pp. 403–57. New York: Wiley.

Bodard, L. 1972. *Green Hell: Massacre of the Brazilian Indians*. New York: Outerbridge and Dienstfrey.

Bolin, B. 1977. Changes of land biota and their importance for the carbon cycle. *Science* 196:613–15.

Bongaarts, J. 1980. Does malnutrition affect fecundity? a summary of evidence. *Science* 208:564–69.

Borgstrom, G. 1965. *The Hungry Planet: The Modern World at the Edge of Famine*. London: Collier-MacMillan.

Bourne, R. 1978. *Assault on the Amazon*. London: Victor Gollancz.

Braga, P. I. S. 1979. Subdivisão fitogeográfica, tipos de vegetação, conservação e inventário da floresta amazônica. *Acta Amazonica* 9(4) suplemento:53–80.

Brazil, ACAR-PARÁ. 1973. Ministério da Agricultura, Associação de Crédito e Assistência Rural do Estado do Pará. Orçamento para 1.000 pés de pimenta-do-reino. Agropolis Brasil Novo: ACAR-PARÁ.

Brazil, ACAR-PARÁ. 1974a. Ministério da Agricultura, Associação de Crédito e Assistência Rural do Estado do Pará. Curso de atualização técnica pedagógica Convênio UNICEF/PIPMO/ABCAR-ACAR-PARÁ. Diciplina: Nutrição. Agrópolis Brasil Novo: ACAR-PARÁ.

Brazil, ACAR-PARÁ. 1974b. Ministério da Agricultura, Associação de Crédito e Assistência Rural do Estado do Pará, Unidade Operacional Altamira IV. Pimenta-do-reino. Agrópolis Brasil Novo: ACAR-PARÁ.

Brazil, ACAR-PARÁ. n.d.[a] (c. 1974). Ministério da Agricultura, Associação de Crédito e Assistência Rural do Estado do Pará. Composição dos alimentos mais comuns em nosso pais. Ceplan/Assessoria Técnica de Alimentação e Habitação, ACAR-PARÁ Curso de Atualização Técnica Pedagógica, Convênio UNICEF/PIPMO/ABCAR/ACAR-PARÁ. Diciplina; Nutrição.

Brazil, ACAR-PARÁ. n.d.[b] (c. 1974). Ministério da Agricultura, Associação de Crédito e Assistência Rural do Estado do Pará. Unidade Operacional Altamira VI. Orçamento para 1.000 pés de cacau. Altamira: ACAR-PARÁ.

Brazil, CEPA-RO. 1980. Governo do Território Federal de Rondônia, Comissão de Planejamento Agrícola de Rondônia. *Perspectiva anual de produção e abastecimento do Território Federal de Rondônia 1979/1980.* Porto Velho: CEPA-RO.

Brazil, DNPEA. 1973a. Ministério da Agricultura, Divisão de Pesquisa Pedológica. *Levantamento de Reconhecimento dos Solos de uma Área Prioritária na Rodovia Transamazônica entre Altamira e Itaituba.* Boletim Técnico No. 34. Rio de Janeiro: DNPEA.

Brazil, DNPEA. 1973b. Ministério da Agricultura, Divisão de Pesquisas Pedológica. *Estudo Expedito dos Solos no Trecho Itaituba-Estreito da Rodovia Transamazônica para Fins de Classificação e Correlação,* Boletim Técnico No. 31. Rio de Janeiro: DNPEA.

Brazil, EMBRAPA-IPEAN. 1974. Ministério da Agricultura, Empresa Brasileira de Pesquisa Agropecuária/Instituto de Pesquisa Agropecuárias do Norte, *Solos da Rodovia Transamazônica; Trecho Itaituba-Rio Branco.* Relatório Preliminar. Belém: EMBRAPA-IPEAN.

Brazil, IBDF. 1975. Ministério da Agricultura, Instituto Brasileiro de Desenvolvimento Florestal. *Inventário Florestal da Rodovia Transamazônica.* Belém: IBDF, Delegacia Estadual do Pará.

Brazil, IBDF. 1979. Ministério da Agricultura, Instituto Brasileiro de Desenvolvimento Florestal. *Plano de Sistema de Unidades de Conservação do Brasil.* Brasília: IBDF.

Brazil, IBDF. 1983. Ministério da Agricultura, Instituto Brasileiro de Desenvolvimento Florestal. *Desenvolvimento Florestal no Brasil.* IBDF Folha Informativa No. 5 (Proj. PNUD/FAO/BRA-82-008). Brasília: IBDF.

Brazil, IBGE. 1978. Presidência da República, Instituto Brasileiro de Geografia e Estatística. WAC Carta Aeronáutica Mundial. Belém, Brasil WAC 2946, escala 1:1.000.000 1ª edição. Rio de Janeiro: IBGE.

Brazil, IBGE. 1980. Presidência da República, Instituto Brasileiro de Geografia e Estatística. *Anuário Estatístico do Brasil 1979,* vol. 40. Rio de Janeiro: IBGE.

Brazil, IBGE. 1982. Presidência da República, Instituto Brasileiro de Geografia e Estatística. *Anuário Estatístico do Brasil 1981,* vol. 42. Rio de Janeiro: IBGE.

Brazil, INCRA. 1972a. Ministério da Agricultura, Instituto Nacional de

Colonização e Reforma Agrária. *Projeto Integrado de Colonização Altamira-I*. Brasília: INCRA.
Brazil, INCRA. 1972b. Ministério da Agricultura, Instituto Nacional de Colonização e Reforma Agrária. *A Colonização no Brasil: Situação Atual, Projeções e Tendências em Rondônia*. Brasília: INCRA.
Brazil, INCRA. n.d. (c. 1972). Ministério da Agricultura, Instituto Nacional de Colonização e Reforma Agrária. *Para um Brasil Redescoberto, Integrado, Nôvo, Forte e Maior, a Transamazônica*. Brasília: INCRA.
Brazil, INCRA. 1974. Ministério da Agricultura, Instituto Nacional de Colonização e Reforma Agrária, *Relatório de Atividades 1974*. Belém: INCRA, Coordenaria Regional do Norte Cr-01.
Brazil, INCRA. 1980. Ministério da Agricultura, Instituto Nacional de Colonização e Reforma Agrária. *Imposto Territorial Rural, Manual de Orientação*. Brasília: INCRA.
Brazil, INPA. 1978. Presidência da República, Instituto Nacional de Pesquisas da Amazônia. *Relatório Quadrianual 1974–1978*. Manaus: INPA.
Brazil, INPA. 1979. Presidência da República, Instituto Nacional de Pesquisas da Amazônia. *Estratégias para a Política Florestal na Amazônia Brasileira*. Acta Amazonica 9(4):1–216 (suplemento).
Brazil, IPEAN. 1966. Ministério da Agricultura, Instituto de Pesquisas Agropecuárias do Norte. Sugestões para adubação (1966), 2a. aproximação. Belém: IPEAN.
Brazil, IPEAN. 1967. Ministério da Agricultura, Instituto de Pesquisas Agropecuárias do Norte. *Contribuição ao Estudo dos Solos de Altamira*, IPEAN Circular No. 10.
Brazil, RADAMBRASIL. Ministério das Minas e Energia, Projeto RADAMBRASIL. 1973–1982. *Levantamento de Recursos Naturais*. vols. 1–23. Rio de Janeiro: Departamento de Produção Mineral.
Brazil, SAGRI e CEPLAC. n.d. (c. 1974). Convênio Banco do Brasil. Roteiro de orçamento para aplicação de credito em lavoura de cacau—1 ha. método derrubada total. Mimeo. Altamira: ACAR-PARÁ.
Brazil, SEMA. 1977. Ministério do Interior, Secretaria Especial do Meio-Ambiente. *Program of Ecological Stations*. Brasília: SEMA.
Brazil, SUDAM. 1978. Ministério do Interior, Superintendência do Desenvolvimento da Amazônia. Estudo da viabilidade técnico-econômico da exploração mecanizada em floresta de terra firme, região de Curuá-Una. PNUD/FAO/IBDF/BRA-76/027. Belém: SUDAM.

Bremner, J. M. and A. M. Blackmer. 1978. Nitrous oxide emission from soils during nitrification of fertilizer nitrogen. *Science* 19:295–96.

Brookfield, H. C. and P. Brown. 1963. *Struggle for Land: Agriculture and Group Territories among the Chimbu of the New Guinea Highlands*. New York: Oxford University Press.

Brooks, E., R. Fuerst, J. Hemming, and F. Huxley. 1973. *Tribes of the Amazon Basin in Brazil 1972*. Report for the Aborigines Protection Society. London: C. Knight.

Brown, H. 1954. *The Challenge of Man's Future*. New York: Viking.

Brown, L. R. 1974. *In the Human Interest: A Strategy to Stabilize World Population*. New York: Norton.

Brown, L. R. 1980. *Building a Sustainable Society*. New York: Norton.

Brush, S. B. 1975. The concept of carrying capacity for systems of shifting cultivation. *American Anthropologist* 77:799–811.

Brush, S. B. 1976. Reply to Vayda. *American Anthropologist* 78:646–47.

Budowski, G. 1956. Tropical savannas, a sequence of forest felling and repeated burnings. *Turrialba* 6:23–33.

Budowski, G. 1976. Why save tropical rain forests? Some arguments for campaigning conservationists. *Amazoniana* 4:529–38.

Budyko, M. I. 1969. The effect of solar radiation variations on the climate of the earth. *Tellus* 21:611–19.

Bunker, S. G. 1979. Power structures and exchange between government agencies in the expansion of the agricultural sector. *Studies in Comparative International Development* 14:56–76.

Bunker, S. G. 1980a. Forces of destruction in Amazonia. *Environment* 22:14–43.

Bunker, S. G. 1980b. Barreiras Burocráticas e institucionais à modernização: o caso da Amazônia. *Pesquisa e Planejamento Econômico* 10:555–600.

Butz, W. P. and J. P. Habicht. 1976. The effects of nutrition and health on fertility: Hypotheses, evidence, and interventions. In R. G. Ridker, ed., *Population and Development: The Search for Selective Interventions*, pp. 210–38. Baltimore: Johns Hopkins University Press.

Calkins, J., ed. 1982. *The Role of Solar Ultraviolet Radiation in Marine Ecosystems*. New York: Plenum.

Camargo, M. N. and I. C. Falesi. 1975. Soils of the Central Plateau and Transamazonic Highway of Brazil. In E. Bornemsza and A. Alvarado, eds., *Soil Management in Tropical America: Proceedings of a Seminar Held at CIAT, Cali, Colombia February 10–14, 1974*, pp. 25–45.

Raleigh: North Carolina State University Soil Science Department.

Cardoso, F. H. and G. Müller. 1978. *Amazônia: Expansão do Capitalismo*, 2d ed. São Paulo: Centro Brasileiro de Análise e Planejamento (CEBRAP), Editora Brasiliense.

Carneiro, R. L. 1960. Slash-and-burn agriculture: A closer look at its implications for settlement patterns. In F. C. Wallace, ed., *Men and Cultures: Selected Papers of the Fifth International Congress of Anthropological and Ethnological Sciences, September 1956*, pp. 229–34. Philadelphia: University of Pennsylvania Press.

Carol, H. 1973. The calculation of theoretical feeding capacity for tropical Africa. *Geographischen Zeitschrift* 61(2):81–94.

Carvalho, J. C. de M. 1981. The conservation of nature and natural resources in the Brazilian Amazon. *CVRD Revista* 2(special ed.):5–47.

Catani, R. A. and A. O. Jacintho. 1974. *Avaliação da Fertilidade do Solo: Método de Análise*. Piracicaba, São Paulo: Livroceres.

Chatt, E. M. 1953. *Cocoa: Cultivation, Processing, Analysis*. New York: Interscience Publishers.

C. e C. (Ciência e Cultura). 1981. Invasão do Parque das Trombetas. *Ciência e Cultura* 33(11):1504.

C. e C. 1982. Ambiente ameaçado. 35(9):1236–37.

C. e C. 1983a. Adeus, Parque! 35(6):834–36.

C. e C. 1983b. Duas novas reservas. 35(2):248.

Clark, C. B. 1973. The economics of overexploitation. *Science* 181:630–34.

Clark, C. B. 1976. *Mathematical Bioeconomics: the Optimal Management of Renewable Resources*. New York: Wiley-Interscience.

Clarke, W. C. 1976. Maintenance of agriculture and human habitats within the tropical forest ecosystem. *Human Ecology* 4:247–59.

Clarke, W. C. 1978. Progressing with the past: Environmentally sustainable modifications of traditional agricultural systems. In E. K. Fisk, ed., *The Adaptation of Traditional Agriculture: Socioeconomic Problems of Urbanization*, pp. 142–57. Development Studies Centre Monograph No. 11. Melbourne: Australian National University.

Clay, J. W. n.d. (1983). The POLONOROESTE Project. In D. Maybury-Lewis, ed., "In the Path of POLONOROESTE: Endangered Peoples of Western Brazil" (typescript), pp. 9–22. Cambridge, Mass.: Cultural Survival.

Coale, A. J. 1983. Recent trends in fertility in less-developed countries. *Science* 221:828–32.

Cochrane, T. T. and P. A. Sánchez. 1982. Land resources, soils, and their management in the Amazon region: A state of knowledge report. In S. B. Hecht, ed., *Amazonia: Agriculture and Land Use Research*, pp. 137–209. Cali, Colombia: Centro Internacional de Agricultura Tropical (CIAT).

Coelho, F. S. and F. Verlengia. 1972. *Fertilidade do Solo*. Campinas, São Paulo: Instituto Campineiro de Ensino Agrícola.

Conklin, H. C. 1959. Population-land balance under systems of tropical forest agriculture. *Proceedings of the Ninth Pacific Science Congress* (Bangkok, 1957), 7:63.

Cooke, G. W. 1970. The carrying capacity of the land in the year 2000. In L. R. Taylor, ed., *The Optimum Population for Britain*, pp. 15–42. New York: Academic Press.

Costa, A. da S., D. A. O. Frazão, E. Tourinho Filho, and A.R.F. Daguer. 1973. *Cultura do Cacau*. Belém: Instituto de Pesquisas Agropecuárias do Norte (IPEAN)/Associação de Crédito e Assistência Rural do Estado do Pará (ACAR-PARÁ), Circular No. 18.

Cowgill, G. L. 1975. On causes and consequences of ancient and modern population changes. *American Anthropologist* 77:505–25.

A *Crítica* [Manaus] December 23, 1978. " 'Contrato de risco' acaba com 40 por cento da floresta amazônica," Section 1, p. 3.

A *Crítica* [Manaus] March 12, 1980. "Inverno não parou a migração para Rondônia," Section 1, p. 7.

A *Crítica* [Manaus] June 1, 1982. "Figueiredo criou ontem 4 estações ecológicas." Section 1, p. 7.

A *Crítica* [Manaus] September 24, 1982. "Produção de borracha chega a 34 mil toneladas este ano." Section 1, p. 7.

Crutzen, P. J., L. E. Heidt, K. P. Krasnec, W. H. Pollock, and W. Seiler. 1979. Biomass burning as a source of atmospheric gases CO, H_2, N_2O, CH_3Cl, and CO_2. *Nature* 282:626.

Cruz, E. de S., G. F. de Souza, and J. B. Bastos. 1971. Influência de adubação NPK no milho, em terra roxa estruturada (Altamira-Zona do Rio Xingú). *Instituto de Pesquisas Agropecuárias do Norte (IPEAN), Série: Fertilidade de Solos* 1(3):1–17.

Cunningham, R. H. 1963. The effect of clearing a tropical forest soil. *Journal of Soil Science* 14:334–44.

da Cunha Camargo, J. G. 1973. *Urbanismo Rural*. Brasilia: Instituto Nacional de Colonização e Reforma Agrária (INCRA).

Dakwa, J. T. 1974. The development of blackpod disease *(Phytophthora palmivora)* in Ghana. *Turrialba* 24:367–72.

da Mata, M., E. W. R. de Carvalho, and M. T. I. I. de C. Silva. 1973. *Migrações Internas no Brasil, Aspectos Econômicos e Demográficos*. Instituto de Planejamento Econômico e Social (IPEA)/Instituto de Pesquisas (INPES) Relatório de Pesquisa No. 19. Rio de Janeiro: IPEA/INPES.

Dantas, M. 1979. Pastagens da Amazônia Central: Ecologia e fauna do solo. *Acta Amazonica* 9(2):1–54 (suplemento).

Dasmann, R. F. 1972. Discussion. In M. T. Farver and J. P. Milton, eds., *The Careless Technology: Ecology and International Development*, pp. 788–89. Garden City, N.Y.: Natural History Press.

Davis, S. 1977. *Victims of the Miracle: Development and the Indians of Brazil*. Cambridge: Cambridge University Press.

de Albuquerque, F. C., and J. M. P. Condurú. 1971. Cultura da pimenta do reino na Região Amazônica. *Instituto de Pesquisas Agropecuárias do Norte (IPEAN) Série: Fitotecnica* 2:1–149.

de Albuquerque, F. C., M. de L. R. Duarte, H. M. Silva, and R. H. M. Pereira. 1973. *A Cultura da Pimenta do Reino*. Belém: Instituto de Pesquisas Agropecuárias do Norte (IPEAN)/Associação de Crédito e Assistência Rural do Estado do Pará (ACAR-PARÁ), Circular No. 19.

de Almeida, H. 1977. *Amazônia*. Belém: Superintendência do Desenvolvimento da Amazônia (SUDAM).

de Almeida, H. 1978. *O Desenvolvimento da Amazônia e a Política de Incentivos Fiscais*. Belém: Superintendência do Desenvolvimento da Amazônia (SUDAM).

de Arruda, H. P. 1972. Exposição do Delegado do Brasil. In Brazil, Ministério da Agricultura, Instituto Nacional de Colonização e Reforma Agrária (INCRA)/Instituto Interamericano de Ciências Agricola da Organização de Estados Americanos (IICA-TRÓPICOS), *Seminário sobre Sistemas de Colonização na Amazônia (Trópico Úmido). Relatório Preliminar*, pp. 5.4–5.9. Belém: IICA-TRÓPICOS.

de Camargo, F. C. 1948. Land and settlement on the recent and ancient quaternary along the railway line of Bragança, State of Pará, Brasil. In *Proceedings of the Inter-American Conference on Conservation of Renewable Natural Resources*, pp. 213–21. Washington, D.C.: U.S. Department of State.

de Carvalho, J. D. n.d. (c. 1974). Ração alimentar. Ministério da Agricultura, Associação de Crédito e Assistencia Rural do Estado do Pará (ACAR-PARÁ).

de Fautereau, E. 1952. *Etudes d'Ecologie Humaine dans l'Aire Amazonienne*. Vendée, France: Fontenay-LeComte.

Denevan, W. M. 1970. The aboriginal population of western Amazonia in relation to habitat and subsistence. *Revista Geografica* 72:61–86.

de Oliveira, A. E., R. Cortez, L. H. Velthem, M. J. Brabo, I. Alves, L. Furtado, I. M. da Silveira, and I. Rodrigues. 1979. Antropologia social e a politica florestal para a Amazônia. *Acta Amazonica* 9(4):191–95 (suplemento).

Dickenson, R. E. 1981. Effects of tropical deforestation on climate. *Studies in Third World Societies* 14:411–41.

do Nascimento, C. N. B. and M. O. D. de Moura Carvalho. 1973. Informações de aspectos pecuários do trópico úmido brasileiro. In *Reunión Técnica de Programación sobre Desarrollo Ganadero de Trópico Húmedo Americano. Guayaquil Ecuador. December 10–14, 1973*. Informe de Conferencias, Cursos y Reuniones No. 30. pp. III-B-1–III-B-57. Belém: Instituto Interamericano de Ciências Agricolas (IICA-TRÓPICOS).

Donahue, T. M. 1975. The SST and ozone depletion. *Science* 187:145.

dos Santos, A.P., E. M. L. de Moraes Novo, and V. Duarte. 1979. *Relatório Final do Projeto INPE/SUDAM*. Relatório No. INPE-1610-RPE/085. São José dos Campos, São Paulo: Instituto Nacional de Pesquisas Espaciais (INPE).

dos Santos, A. P., E. M. L. de Moraes Novo, and V. Duarte. 1980. *Exemplo de Aplicação de Dados do Sistema LANDSAT, no Estudo das Relações entre Compartimentação Topográfica e Qualidade de Pastagens no Município de Paragominas (PA)*. Relatório No. INPE-1756-RPE/145. São José dos Campos, São Paulo: Instituto Nacional de Pesquisas Espaciais (INPE).

Draper, N. R. and H. Smith. 1966. *Applied Regression Analysis*. New York: Wiley.

Durham, W. E. 1976. Resource competition and human aggression, Part I: A review of primitive war. *Quarterly Review of Biology* 51:385–415.

Dynia, J. F., G. N. C. Moreira, and R. M. Bloise. 1977. Fertilidade de solos da região da Rodovia Transamazônica. II. Fixação do fósforo em podzólico vermelho-amarelo e terra roxa estruturada latossólica. *Pesquisa Agropecuária Brasileira* 12:75–80.

Eckholm, E. 1978. Disappearing species: The social challenge. Worldwatch Paper No. 22. Washington, D.C.: Worldwatch Institute.

Eden, M. J. 1978. Ecology and land development: the case of Amazonian

rainforest. *Transactions of the Institute of British Geographers, New Series.* 3(4):444–63.
Egler, E. G. 1961. A Zona Bragantina do Estado do Pará. *Revista Brasileira de Georgrafia* 23(3):527–55.
Ehrlich, P. R. 1982. Human carrying capacity, extinctions, and nature reserves. *BioScience* 32(5):331–33.
Ehrlich, P. R. and A. H. Ehrlich. 1981. *Extinction: The Causes and Consequences of the Disappearance of Species.* New York: Random House.
Ehrlich, P. R., A. H. Ehrlich, and J. P. Holdren. 1977. *Ecoscience: Population, Resources, Environment.* San Francisco: Freeman.
Eigner, J. 1975. Unshielding the sun: Environmental effects. *Environment* 17(3):15–25.
O Estado de São Paulo. May 24, 1974. "Ocupação tornerá Amazônia rentável," p. 11.
O Estado de São Paulo. May 21, 1976. "Cotrijuí inicia no próximo mês a instalação de núcleo na Amazônia," p. 12.
Faechem, R. 1973. A clarification of carrying capacity formulae. *Australian Geographical Studies* 11:234–36.
Falesi, I. C. 1967. O estado atual dos conhecimentos sobre os solos da Amazônia Brasileira. In H. Lent, ed., *Atas do Simpósio sobre a Biota Amazônica,* 1:151–68. Rio de Janeiro: Conselho Nacional de Pesquisas.
Falesi, I. C. 1972a. *Solos da Rodovia Transamazônica.* Boletim Técnico No. 55. Belém: Instituto de Pesquisas Agropecuárias do Norte (IPEAN).
Falesi, I. C. 1972b. O estado atual dos conhecimentos sobre os solos da Amazônia Brasileira. Parte I. In *Zoneamento Agrícola da Amazônia (1a. Aproximação).* Boletim Técnico No. 54, pp. 17–67. Belém: Instituto de Pesquisas Agropecuárias do Norte (IPEAN).
Falesi, I. C. 1974a. Soils of the Brazilian Amazon. In C. Wagley, ed., *Man in the Amazon,* pp. 201–29. Gainesville: University Presses of Florida.
Falesi, I. C. 1974b. O solo na Amazônia e sua relação com a definição de sistemas de produção agrícola. In Empresa Brasileira de Pesquisas Agropecuárias (EMBRAPA). *Reunião do Grupo Interdisciplinar de Trabalho sobre Diretrizes de Pesquisa Agrícola para a Amazônia (Trópico Úmido). Brasília, Maio 6–10, 1974,* 1:2.1–2.11. Brasília: EMBRAPA.
Falesi, I. C. 1976. *Ecossistema de Pastagem Cultivada na Amazônia Brasileira.* Boletim Técnico No. 1. Belém: Centro de Pesquisa Agropecuária do Trópico Úmido (CPATU).

Falesi, I. C. and T. E. Rodriques. 1970. As terras roxas na Amazônia Brasileria. *Caderno de Ciências da Terra.* 6:1–19.
Farnworth, E. G. and F. B. Golley, eds. 1974. *Fragile Ecosystems: Evaluation of Research and Applications in the Neotropics.* New York: Springer.
Fearnside, P. M. 1972. An estimate of carrying capacity of the Osa Penninsula for human populations supported on a shifting agriculture technology. In Organization for Tropical Studies (OTS), *Report of Research Activities Undertaken during the Summer of 1972*, pp. 486–52. San José, Costa Rica: OTS.
Fearnside, P. M. 1974. Preliminary models for estimation of carrying capacity for human populations in a colonization area of the Transamazon Highway, Brazil. Mimeo.
Fearnside, P. M. 1978. *Estimation of Carrying Capacity for Human Populations in a Part of the Transamazon Highway Colonization Area of Brazil.* Ph.D. dissertation, University of Michigan. Ann Arbor: University Microfilms International.
Fearnside, P. M. 1979a. Cattle yield prediction for the Transamazon Highway of Brazil. *Interciencia* 4:220–25.
Fearnside, P. M. 1979b. The development of the Amazon rain forest: priority problems for the formulation of guidelines. *Interciencia* 4:338–43.
Fearnside, P. M. 1979c. O processo de desertificação e os riscos de sua ocorrencia no Brasil. *Acta Amazonica* 9:393–400.
Fearnside, P. M. 1979d. O agro-ecossistema dos colonos da Transamazônica: simulação de produções de milho. *Ciência e Cultura* 31(7):414 (suplemento).
Fearnside, P. M. 1979e. *The Simulation of Carrying Capacity for Human Agricultural Populations in the Humid Tropics: Program and Documentation.* Manaus: Instituto Nacional de Pesquisas da Amazônia (INPA).
Fearnside, P. M. 1980a. Black pepper yield prediction for the Transamazon Highway of Brazil. *Turrialba* 30:35–42.
Fearnside, P. M. 1980b. Land-use allocation of the Transamazon Highway colonists of Brazil and its relation to human carrying capacity. In F. Barbira-Scazzocchio, ed., *Land, People, and Planning in Contemporary Amazonia,* Occasional Paper No. 3, pp. 114–38. Cambridge, U.K.: Cambridge University, Centre of Latin-American Studies.
Fearnside, P. M. 1980c. The effects of cattle pastures on soil fertility in the Brazilian Amazon: Consequences for beef production sustainability. *Tropical Ecology* 21:125–37.

Fearnside, P. M. 1980d. The prediction of soil erosion losses under various land uses in the Transamazon Highway Colonization Area of Brazil. In J. I. Furtado, ed., *Tropical Ecology and Development: Proceedings of the Fifth International Symposium of Tropical Ecology, April 16–21, 1979, Kuala Lumpur, Malaysia*, pp. 1287–95. Kuala Lumpur: International Society for Tropical Ecology (ISTE).

Fearnside, P. M. 1982. Deforestation in the Brazilian Amazon: How fast is it occurring? *Interciencia* 7(2):82–88.

Fearnside, P. M. 1983a. Land use trends in the Brazilian Amazon region as factors in accelerating deforestation. *Environmental Conservation* 10(2):141–48.

Fearnside, P. M. 1983b. Stochastic modeling and human carrying capacity estimation: a tool for development planning in Amazonia. In E. F. Moran, ed., *The Dilemma of Amazonian Development*, pp. 279–95. Boulder, Colo.: Westview Press.

Fearnside, P. M. 1983c. Development alternatives in the Brazilian Amazon: an ecological evaluation. *Interciencia* 8(2):65–78.

Fearnside, P. M. 1984a. A floresta vai acabar? *Ciência Hoje* 2(10):42–52.

Fearnside, P. M. 1984b. Previsão de produções de cacau na rodovia Transamazônica. *Ciência e Cultura*, vol. 36, no. 7, suplemento: 6 (Abstract).

Fearnside, P. M. 1984c. Brazil's Amazon settlement schemes: Conflicting objectives and human carrying capacity. *Habitat International* 8(1):45–61.

Fearnside, P. M. 1984d. Ecological research reserve for Brazil's Amazon rainforest established in Ouro Preto do Oeste, Rondônia. *Environmental Conservation* 11(3):273–74.

Fearnside, P. M. 1984e. Initial soil quality conditions on the Transamazon Highway of Brazil and their simulation in models for estimating human carrying capacity. *Tropical Ecology*, vol. 25, no. 2.

Fearnside, P. M. 1984f. Simulation of meteorological parameters for estimating human carrying capacity in Brazil's Transamazon Highway colonization area. *Tropical Ecology* 25(1):136–44.

Fearnside, P. M. 1984g. Land clearing behavior in small farmer settlement schemes in the Brazilian Amazon and its relation to human carrying capacity. In A. C. Chadwick and S. L. Sutton, eds., *Tropical Rain Forest: Ecology and Management*, pp. 255–71. Leeds, U.K.: Leeds Philosophical and Literary Society.

Fearnside, P. M. 1985a. Brazil's Amazon forest and the global carbon problem. *Interciencia*, vol. 10, no. 4.

Fearnside, P. M. 1985b. Environmental change and deforestation in the Brazilian Amazon. In J. Hemming, ed., *Change in the Amazon Basin: Man's Impact on Forest and Rivers*, pp. 71–89. Manchester, U.K.: Manchester University Press.

Fearnside, P. M. n.d.(a). Deforestation and decision-making in the development of Brazilian Amazonia. Paper presented at the International Symposium on Amazonia, July 7–13, 1983. Belém, Pará. *Interciencia* (forthcoming).

Fearnside, P. M. n.d.(b). Agriculture in Amazonia. In G. T. Prance and T. E. Lovejoy, eds., *The Amazon Rain Forest*, pp. 393–418. Oxford, U.K.: Pergamon Press. (In press).

Fearnside, P. M. n.d.(c). Burn quality prediction for simulation of the agricultural system of Brazil's Transamazon Highway colonists for estimating human carrying capacity. In K. S. Misra, H. V. Pandey, and G. S. Govil, eds., *Ecology and Resource Management in the Tropics*. Varanasi, India: International Society for Tropical Ecology (ISTE). (In press).

Fearnside, P. M. n.d.(d). *Data Management Package for Carrying Capacity Estimation in the Humid Tropics*. (In preparation).

Fearnside, P. M. n.d.(e). A stochastic model for estimating human carrying capacity in Brazil's Transamazon Highway colonization area. *Human Ecology* (in press).

Fearnside, P. M. and G. de L. Ferreira. 1984. Roads in Rondônia: highway construction and the farce of unprotected reserves in Brazil's Amazonian forest. *Environmental Conservation* 11(4):358–60.

Fearnside, P. M. and J. M. Rankin. 1980. Jari and development in the Brazilian Amazon. *Interciencia* 5:146–56.

Fearnside, P. M. and J. M. Rankin. 1982a. The new Jari: Risks and prospects of a major Amazonian development. *Interciencia* 7(6):329–39.

Fearnside, P. M. and J. M. Rankin. 1982b. Jari and Carajás: The uncertain future of large silvicultural plantations in the Amazon. *Interciencia* 7(6):326–28.

Fearnside, P. M. and J. M. Rankin. 1985. Jari revisited: changes and the outlook for sustainability in Amazonia's largest silvicultural estate. *Interciencia* 10(3):121–29.

Fife, D. 1971. Killing the goose. *Environment* 13(3):20–27.

Fisher, R. A. 1958. *The Genetical Theory of Natural Selection*. New York: Dover.

Fittkau, E. J. and H. Klinge. 1973. On biomass and trophic structure

Central Amazonian rain forest ecosystem. *Biotropica* 5:2–14.
Flegg, A. T. 1979. The role of inequality of income in the determination of birth rates. *Population Studies* 33:457–77.
Fleming-Moran, M. and E. F. Moran. 1978. O surgimento de classes sociais numa communidade planejada para ser igualitária. *Boletim do Museu Paraense Emílio Goeldi: Nova Série (Antropologia)*, No. 69.
Flohn, H. 1974. Climatic variation and modification of climate: facts and problems. *Applied Sciences and Development* 8:96–105.
Fonsêca, R., A. C. Dias, A. Pinho, E. Pires, E. Miranda, P. Cabola, and C. Santana. 1969. Correlações dos teores de fósforo nos solos com respostas de microparcelas de milho na zona cacaueira da Bahía. In L. C. Cruz, ed., *Memórias da Sequnda Conferência Internacional de Pesquisas em Cacau. November 19–26, 1967, Salvador e Itabuna, Bahía, Brasil*, pp. 487–97. Itabuna: Comissão Executiva do Plano da Lavoura Cacaueira (CEPLAC).
Forrester, J. W. 1970. *Principles of Systems*. Cambridge, Mass.: Wright-Allen Press.
Forrester, J. W. 1971. *World Dynamics*. Cambridge, Mass.: Wright-Allen Press.
Found, W. C. 1971. *A Theoretical Approach to Rural Land Use Patterns*. London: Edward Arnold.
Fox, D. J. and K. E. Guire. 1976. *Documentation for MIDAS. 3rd. ed. September 1976*. Ann Arbor: University of Michigan, Statistical Research Laboratory.
Fox, D. L., R. Kamens, and H. E. Jeffries. 1975. Stratospheric nitric oxide: Measurements during daytime and sunset. *Science* 188:1111–13.
Frank, P. W. 1957. Coactions in laboratory populations of two species of *Daphnia*. *Ecology* 38:510–19.
Freeman, J. D. 1955. *Iban Agriculture, a Report on the Shifting Cultivation of Hill Rice by the Iban of Sarawak*, vol. 18. London: Colonial Research Studies.
Freise, F. W. 1934. Beobachtungen über den Verbleib von Niederschlägen im Urwald und der Einfluss von Valdbestand auf den Wasserlhaushalt der Umgebung. *Forstwissentschaftliches Centralblaat* 56:231–45.
Freise, F. W. 1939. Untersuchungen über die Folgen der Brandwirtschaft aus tropischen Boden. *Tropenpflanzer* 42:1–22.
Frota Neto. 1978. Plano: vender a floresta Amazônica. *Folha de São Paulo*. December 3, 1978, Section 1, pp. 1–2.
Glassow, M. A. 1978. The concept of carrying capacity in the study of cul-

ture processes. In M. B. Schiffer, ed., *Advances in Archaeological Method and Theory*, 1:31-48. New York: Academic Press.

Gómez-Pompa, A., C. Vásquez-Yanes, and S. Gueriara. 1972. The tropical rain forest: A nonrenewable resource. *Science* 177:762-65.

Gonçalves, J. R. C. 1970. Recentes pesquisas sôbre doenças da seringueira. *Instituto de Pesquisas Agropecuárias do Norte (IPEAN) Série: Fitotecnia*, vol. 1, no. 4.

Goodland, R. J. A. 1980a. Environmental ranking of Amazonian development projects. In F. Barbira-Scazzocchio, ed., *Land, People, and Planning in Contemporary Amazonia*. Occasional Paper No. 3, pp. 1–20. Cambridge, U.K.: Cambridge University, Centre of Latin-American Studies.

Goodland, R. J. A. 1980b. Environmental ranking of Amazonian development projects in Brazil. *Environmental Conservation* 7:9-26.

Goodland, R. J. A. and H. S. Irwin. 1975a. *Amazon Jungle: Green Hell to Red Desert? an Ecological Discussion of the Environmental Impact of the Highway Construction Program in the Amazon Basin*. New York: Elsevier.

Goodland, R. J. A. and H. S. Irwin. 1975b. *A Selva Amazônica: do Inferno Verde ao Deserto Vermelho?* Tradução de R. R. Jungueira; revisão técnica, prefácio e notas de M. G. Ferri. São Paulo: Ed. Itaiaia/Ed. da Universidade de São Paulo.

Goodland, R. J. A. and H. S. Irwin. 1977. Amazonian forest and cerrado: development and environmental conservation. In G. T. Prance and T. S. Elias, eds., *Extinction Is Forever*, pp. 214-33. Bronx: New York Botanical Garden.

Goodland, R. J. A., H. S. Irwin, and G. Tillman. 1978. Ecological development for Amazonia. *Ciência e Cultura* 30:275-89.

Goreau, T. J. 1981. *Biogeochemistry of Nitrous Oxide*. Ph.D. dissertation, Harvard University, Cambridge, Mass.

Goreau, T. J., W. A. Kaplan, S. C. Wofsy, M. B. McElroy, F. W. Valois, and S. W. Watson. 1980. Production of NO_2 and N_2O by nitrifying bacteria at reduced concentrations of oxygen. *Applied Environmental Microbiology* 40:526-32.

Goulding, M. 1980. *The Fishes and the Forest*. Berkeley: University of California Press.

Gourou, P. 1966. *The Tropical World: Its Social and Economic Conditions and Its Future Status*. 4th ed. Trans. by S. H. Beaver and E. D. Tabunde. New York: Longman.

Gourou, P. 1971. Leçons de Geografie Tropicale. Leçons Données a le College de France de 1947 a 1970. Paris: Mouton.
Grainger, A. 1980. The state of the world's tropical forests. The Ecologist 10(1):6–15.
Graner, E. A., E. W. L. Orsi, F. F. Toledo, O. P. Godong, J. J. M. Abrahão, and J. D. Costa. 1972. Plantas Alimenticias: Arroz, Feijão. Piracicaba, São Paulo: Departamento de Agricultura e Horticultura, Escola Superior de Agricultura "Luis de Queirroz."
Greenland, D. J. and R. Herrera. 1978. Patterns of use of tropical forest ecosystems 4. Shifting cultivation and other agricultural practices. In F. Dicastri and A. Sasson, eds., State of Knowledge Report on Tropical and Subtropical Forest Ecosystems. Paris: UNESCO.
Greenland, D. J. and P. H. Nye. 1959. Increases in the carbon and nitrogen contents of tropical soils under natural fallows. Journal of Soil Science 10:284–99.
Grobecker, A. J., S. C. Coroniti, and R. H. Cannon, Jr. 1974. The Effects of Stratospheric Pollution by Aircraft: Report of Findings. Executive Summary. Springfield, Va.: National Technical Information Service.
Gross, D. R. 1975. Protein capture and cultural development in the Amazon Basin. American Anthropologist 77:526–49.
Gross, D. R. and B. A. Underwood. 1971. Technological change and caloric costs: Sisal agriculture in northeastern Brazil. American Anthropologist 73:725–40.
Guillemin, R. 1956. Evolution de l'agriculture autochthone dans les savannes de l'Oubangui. Agronomie Tropicale 11(1):39–61; 11(2):143–76; 11(3):279–309.
Guimarães, G. de A., J. B. Bastos, and E. de C. Lopes. 1970. Métodos de análise física, química e instrumental de solos. Instituto de Pesquisas e Experimentação Agropecuárias do Norte (IPEAN) Série: Química de Solos 1(1):1–108.
Hairston, N. G., J. D. Allen, R. K. Colwell, D. J. Futuyma, J. Howell, J. D. Mathias, and J. H. Vandermeer. 1969. The relationship between species diversity and stability: An experimental approach with protozoa and bacteria. Ecology 49:1091–1101.
Hammond, A. L. 1977a. Remote sensing. I. Landsat takes hold in South America. Science 196:511–12.
Hammond, A. L. 1977b. Remote sensing. II. Brazil explores its Amazon wilderness. Science 196:513–15.

Hanbury-Tenison, R. 1973. A *Question of Survival for the Indians of Brazil*. London: Angus and Robertson.
Hardesty, D. L. 1977. *Ecological Anthropology*. New York: Wiley.
Hardin, G. 1968. The tragedy of the commons. *Science* 162:1243–48.
Hardy, F. 1961. *Manual de cacao*. Turrialba, Costa Rica: Instituto Interamericano de Ciencias Agricolas (IICA).
Harris, D. R. 1971. The ecology of swidden cultivation in the upper Orinoco rainforest, Venezuela. *Geographical Review* 61:475–95.
Hayden, B. 1975. The carrying capacity dilemma: An alternative approach. In A. C. Swedlund, ed., *Population Studies in Archaeology and Biological Anthropology: A Symposium*, pp. 11–21. Washington, D.C.: Society for American Archaeology, Memoir 30.
Hecht, S. B. 1981. Deforestation in the Amazon basin: Practice, theory and soil resource effects. *Studies in Third World Societies* 13:61–108.
Hecht, S. B. 1982. Agroforestry in the Amazon basin: practice, theory and limits of a promising land use. In S. B. Hecht, ed., *Amazonia: Agriculture and Land Use Research*, pp. 331–71. Cali, Colombia: Centro Internacional de Agricultura Tropical (CIAT).
Hecht, S. B. 1983. Cattle ranching in the eastern Amazon: Environmental and social implications. In E. F. Moran, ed., *The Dilemma of Amazonian Development*. pp. 155–88. Boulder, Colo.: Westview Press.
Henderson-Sellers, A. and V. Gornitz. 1984. Possible climatic impacts of land cover transformations, with particular emphasis on tropical deforestation. *Climatic Change* 6:231–57.
Herrera, R., C. F. Jordan, H. Klinge, and E. Medina. 1978. Amazon ecosystems: Their structure and functioning with particular emphasis on nutrients. *Interciencia* 3:223–32.
Hirano, C. 1974. Projeto Iriri: estudo dos solos da área. (Unpublished Report) Brasília: Instituto Nacional de Colonização e Reforma Agrária (INCRA).
Hobbie, J., J. Cole, J. Dungan, R. A. Houghton, and B. Peterson. 1984. Role of biota in global CO_2 balance: the controversy. *BioScience* 34(8):492–498.
Homma, A. K. O. 1976. "Programação das Atividades Agropecuárias, sob Condições de Risco, nos Lotes do Núcleo de Colonização de Altamira." Masters thesis in agricultural economics, Universidade Federal de Viçosa, Viçosa, Minas Gerais.
Homma, A. K. O. and L. Miranda Filho. 1979. *Análise da Estrutura da produção de Pimenta-do-Reino no Estado do Pará-1977/78*. Belém:

Empresa Brasileira de Pesquisa Agropecuária (EMBRAPA), Comunicado Técnico No. 20.
Homma, A. K. O., R. M. F. Viégas, J. Graham, J. de J. S. Lemos, and J. C. dos Mendes Lopes. 1978. *Identificação de Sistemas de Produção nos Lotes do Núcleo de Colonização de Altamira, Pará*. Belém: Empresa Brasileira de Pesquisa Agropecuária-Centro de Pesquisa Agropecuária do Trópico Úmido (EMBRAPA-CPATU), Comunicado Técnico No. 4.
Houghton, R. A., J. E. Hobbie, J. M. Melillo, B. Moore, B. J. Peterson, G. R. Shaver, and G. M. Woodwell. 1983. Changes in the carbon content of terrestrial biota and soils between 1860 and 1980: a net release of CO_2 to the atmosphere. *Ecological Monographs* 53(3):235–62.
Hubbell, S. P. 1973. Populations and simple food webs as energy filters. I. One-species systems. *American Naturalist* 107:94–121.
Hunter, J. M. 1966. Ascertaining population carrying capacity under traditional agriculture in developing countries: Note on a method employed in Ghana. *Professional Geographer* 18:151–54.
Ianni, O. 1979. *Colonização e Contra-Reforma Agrária na Amazônia*. Petrópolis, Rio de Janeiro: Editora Vozes.
IBRD (International Bank for Reconstruction and Development). 1981. *Brazil: Integrated Development of the Northwest Frontier*. Washington, D.C.: The World Bank (IBRD).
Idso, S. B. 1980a. The climatological significance of a doubling of earth's atmospheric carbon dioxide concentration. *Science* 207:1462–63.
Idso, S. B. 1980b. Carbon dioxide and climate. *Science* 210:7–8.
IICA-Trópicos (Instituto Interamericano de Ciências Agricolas, Programa Cooperativa para el Trópico Americano). 1972. *Bibliografía sobre Colonización en América Latina*. Turrialba, Costa Rica: IICA-TROPICOS.
Irion, G. 1978. Soil infertility in the Amazonian rain forest. *Naturwissenschaften* 65:515–19.
Jacobs, M. 1980. Significance of the tropical rain forest on 12 points. *BioIndonesia* 7:75–94.
Janzen, D. H. 1970a. Herbivores and the number of tree species in tropical forests. *American Naturalist* 104:501–28.
Janzen, D. H. 1970b. The unexploited tropics. *Ecological Society of America Bulletin* 51:4–7.
Janzen, D. H. 1972a. Whither tropical ecology? In J. A. Behnke, ed., *Challenging Biological Problems*, pp. 281–96. New York: Oxford University Press.

Janzen, D. H. 1972b. The uncertain future of the tropics. *Natural History* 81:80–90.

Janzen, D. H. 1972c. Interfield and interplant spacing in tropical insect control. In *Proceedings of the Annual Tall Timbers Conference on Ecological Animal Control by Habitat Management, February 24–25, 1972*, pp. 1–6.

Janzen, D. H. 1973a. Tropical agroecosystems: Habitats misunderstood by the temperate zones, mismanaged by the tropics. *Science* 182:1212–19.

Janzen, D. H. 1973b. Sweep samples of tropical foliage insects: Effects of seasons, vegetation types, elevation, time of day, and insularity. *Ecology* 54:687–708.

Janzen, D. H. 1974. The deflowering of Central America. *Natural History* 83:48–53.

Janzen, D. H. 1976. Why bamboos wait so long to flower. *Annual Review of Ecology and Systematics* 7:347–91.

Jones, G. W. 1979. Indonesia: The transmigration programme and development planning. In R. J. Pryor, ed., *Migration and Development in South-East Asia: A Demographic Perspective*, pp. 212–21. Kuala Lumpur: Oxford University Press.

Jordan, C. F., R. L. Todd, and E. Escalante. 1979. Nitrogen conservation in a tropical rainforest. *Oecologia* 39:123–28.

Junqueira, C. n.d. (1983). Cinta Larga. In D. Maybury-Lewis, ed., "In the Path of POLONOROESTE: Endangered Peoples of Western Brazil" (typescript), pp. 55–58. Cambridge, Mass.: Cultural Survival.

Katzman, M. T. 1976. Paradoxes of Amazonian development in a "resource-starved" world. *Journal of Developing Areas* 10(4):445–60.

Kerr, R. A. 1982. CO_2-climate models defended. *Science* 217:620.

Kerr, R. A. 1983. Carbon dioxide and a changing climate. *Science* 222:491.

Kingsland, S. 1982. The refractory model: The logistic curve and the history of population ecology. *Quarterly Review of Biology* 57:29–52.

Kleinpenning, J. M. G. 1975. *The Integration and Colonization of the Brazilian Portion of the Amazon Basin*. Nijmegen, Holland: Institute of Geography and Planning.

Kleinpenning, J. M. G. 1979. *An Evaluation of the Brazilian Policy for the Integration of the Amazon Basin (1964–1975)*. Nijmegen, Holland: Vakroep Sociale Geografie van de Ontwikkelinsgslanden, Geografisch en Planologisch Instituut Publikatie 9.

Kohlhepp, G. 1980. Analysis of state and private regional development projects in the Brazilian Amazon basin. *Applied Geography and Development* 16:53–79.

Koster, H. W., E. J. A. Khan, and R. P. Bosshart. 1977. *Programa e Resultados Preliminares dos Estudos de Pastagens na Região de Paragominas, Pará, e nordeste de Mato Grosso junho 1975-dezembro 1976*. Belém: Superintendência do Desenvolvimento da Amazônia (SUDAM), Convênio SUDAM/Instituto de Pesquisas IRI.

Krebs, C. J. 1972. *Ecology: The Experimental Analysis of Distribution and Abundance*. New York: Harper and Row.

Lanly, J. P. and M. Gillis. 1980. *Provisional Results of the FAO/UNEP Tropical Forest Resources Assessment Project, Tropical America*. Rome: UN-FAO.

Lathrap, D. 1968. The "hunting" economies of the tropical forest zone of South America. In R. Lee and I. de Vore, eds., *Man the Hunter*. Chicago: Aldine.

Leopoldo, P. R., W. Franken, and E. Salati. 1982. Balanço hídrico de pequena bacia hidrográfica em floresta amazônica de terra firme. *Acta Amazonica* 12(2):333–37.

Leovy, C. B. 1980. Carbon dioxide and climate. *Science* 210:7.

Levins, R. 1969. The effect of random variations of different types on population growth. *Proceedings of the National Academy of Sciences* 62:1061–65.

Lewontin, R. C. 1970. The units of selection. *Annual Review of Ecology and Systematics* 1:1–18.

Lian, M. S. and R. D. Cess. 1977. Energy balance climate models: A reappraisal of ice-albedo feedback. *Journal of the Atmospheric Sciences* 34:1058–62.

Lima, A. da S. 1973. *La Mise en Valeur des Terres Nouvelles. le Cas de l'Amazonie Bresilienne*. These de 3e cycle. Paris: Université de Paris I, Panthéon, Sorbonne, Paris. Ministère de l'Education Nationale, Ecole Pratique des Hautes Etudes VI Section—Sciences Economicas et Sociales, Centre International de Recherche sur l'Environnment et le Développment, Travaux et Etudes No. 1.

List, R. J. 1958. *Smithsonian Meteorological Tables*. Washington, D.C.: Smithsonian Institution.

Lloyd, W. F. 1969 (1833). Two lectures on the checks to population. In G. Hardin, ed., *Population, Evolution, and Birth Control: A Collage of Controversial Ideas*. 2d ed. San Francisco: Freeman.

Lovejoy, T. E. 1973. The Transamazonica: Highway to extinction? *Frontiers* 1973 (Spring):25–30.

Lovejoy, T. E., R. O. Bierregaard, J. M. Rankin, and H. O. R. Schubart. 1983. Ecological dynamics of tropical forest fragments. In S. L. Sutton, T. C. Whitmore, and A. C. Chadwick, eds., *Tropical Rain Forest: Ecology and Management*, pp. 377–384. Oxford, U.K.: Blackwell Scientific Publications.

McArthur, M. 1977. Nutritional research in Melanesia: A second look at the Tsembaga. In T. P. Bayliss-Smith and R. G. Feachem, eds., *Subsistence and Survival: Rural Ecology in the Pacific*, pp. 91–128. New York: Academic Press.

McCabe, J. L. and M. R. Rosenzweig. 1976. Female employment creation and family size. In R. G. Ridker, ed., *Population and Development: The Search for Selective Interventions*, pp. 322–55. Baltimore: Johns Hopkins University Press.

McElroy, M. B., J. W. Elkins, S. C. Wofsy, and Y. L. Yung. 1976. Sources and sinks for atmospheric N_2O. *Review of Geophysics and Space Research* 14:143.

McGregor, D. F. M. 1980. An investigation of soil erosion in the Colombian rainforest zone. *Catena* 7(4):265–73.

McIntyre, L. 1980. Jari: A massive technology transplant takes root in the Amazon jungle. *National Geographic*, May 1980, pp. 693–711.

Mahar, D. J. 1979. *Frontier Development Policy in Brazil: A Study of Amazonia*. New York: Praeger.

Malthus, T. R. 1965 (1798). *An Essay on the Principle of Population as it Affects the Future Improvement of Society*. In E. J. Kormondy, ed., *Readings in Ecology*, pp. 62–63. Englewood Cliffs, N.J.: Prentice-Hall.

Malthus, T. R. 1960 (1830). A summary view of the principle of population. In T. R. Malthus, J. Huxley, and F. Osborn. *Three Essays on Population* pp. 13–59. New York: Mentor.

Manabe, S. and R. J. Stouffer. 1979. A CO_2-climate sensitivity study with a mathematical model of global climate. *Nature* 282:491–93.

Manabe, S. and R. T. Wetherald. 1967. Thermal equilibrium of the atmosphere with a given distribution of relative humidity. *Journal of the Atmospheric Sciences* 24:241–59.

Manabe, S. and R. T. Wetherald. 1975. The effects of doubling the CO_2 concentration on the climate of a general circulation model. *Journal of the Atmospheric Sciences* 32:3–15.

Mankin, J. B., R. V. O'Neill, H. H. Shugart, and B. W. Rust. 1977. The importance of validation in ecosystem analysis. In G. S. Innis, ed., *New Directions in the Analysis of Ecological Systems*, Part 1, pp. 63–71. Simulation Councils Proceedings Series, vol. 5, La Jolla, Calif.: Society for Computer Simulation.

Margalef, R. 1968. *Perspectives in Ecological Theory*. Chicago: University of Chicago Press.
Marques, J., J. M. dos Santos, N. A. Villa Nova, and E. Salati. 1977. Precipitable water and water vapor flux between Belém and Manaus. *Acta Amazonica* 7:355–62.
Marshall, E. 1981. By flood, if not by fire, CEQ says. *Science* 211:463.
Martine, G. 1979. Colonization in Rondônia and the reproduction of conditions prevailing in older areas. Paper presented at the Informal Workshop on Migration Policies, Geneva, December 1979. UNDP/Human Resources Planning Project BRA/70/55.
Martine, G. 1980. Recent colonization experiences in Brazil: Expectations versus reality. In F. Barbira-Scazzocchio, ed., *Land, People, and Planning in Contemporary Amazonia*. Occasional Paper No. 3, pp. 80–94. Cambridge, U.K.: Cambridge University, Centre of Latin-American Studies.
Martins, J. de S. 1980. Fighting for land: Indians and *posseiros* in Legal Amazonia. In F. Barbira-Scazzocchio, ed., *Land, People, and Planning in Contemporary Amazonia*. Occasional Paper No. 3, pp. 95–105. Cambridge, U.K.: Cambridge University, Centre of Latin-American Studies.
Maugh, T. H. III. 1982. New link between ozone and cancer. *Science* 216:396–97.
May, R. M. 1973. *Stability and Complexity in Model Ecosystems. Monographs in Population Biology* No. 4. Princeton, N.J.: Princeton University Press.
Meadows, D. H., D. L. Meadows, J. Randers, and W. W. Behrens, III. 1972. *The Limits to Growth*. New York: New American Library.
Meadows, D. H., D. L. Meadows, J. Randers, and W. W. Behrens, III. 1973. A response to Sussex. In H. S. D. Cole, C. Freeman, M. Jahoda, and K. L. R. Pavitt, eds., *Models of Doom: A Critique of the Limits to Growth*, pp. 217–40. New York: Universe Books.
Meadows, D. L., W. W. Behrens, III, D. H. Meadows, R. F. Naile, J. Randers, and E. K. O. Zahn. 1973. *The Dynamics of Growth in a Finite World*. Cambridge, Mass.: Wright-Allen Press.
Meggers, B. J. 1971. *Amazonia: Man and Culture in a Counterfeit Paradise*. Chicago: Aldine.
Mercer, J. H. 1978. West Antarctic ice sheet and CO_2 greenhouse effect: A threat of disaster. *Nature* 271:321–25.
Mesarovic, M. and E. Pestle. 1974a. *Mankind at the Turning Point: The*

Brasileira para o Progresso da Ciência, Fortaleza, Ceará, July 11–19, 1979. Mimeo. (Abstract.)

Myers, N. 1976. An expanded approach to the problem of disappearing species. *Science* 193:198–202.

Myers, N. 1979. *The Sinking Ark: A New Look at the Problem of Disappearing Species.* New York: Pergamon.

Myers, N. 1980a. *Conversion of Tropical Moist Forests.* Washington, D.C.: National Academy of Sciences.

Myers, N. 1980b. The present status and future prospects of tropical moist forests. *Environmental Conservation* 7(2):101–14.

Myers, N. 1984. *The Primary Source: Tropical Forests and our Future.* New York: Norton.

National Research Council Committee on Research Priorities in Tropical Biology. 1980. *Research Priorities in Tropical Biology.* Washington, D.C.: National Academy of Sciences Press.

Nelson, M. 1973. *The Development of Tropical Lands: Policy Issues in Latin America.* Baltimore: Johns Hopkins University Press.

Neves, A. M. and A. M. T. Lopes. 1979. Os projetos de colonização. In O. Valverde, ed., A *Organização do Espaço na Faixa da Transamazônica*, pp. 80–120. Rio de Janeiro: Instituto Brasileiro de Geografia e Estatística (IBGE).

Newell, R. E. and T. G. Dopplick. 1979. Questions concerning the possible influence of anthropogenic CO_2 on atmospheric temperature. *Journal of Applied Meteorology* 18:822–25.

Newland, K. 1977. Women and population growth: Choice beyond childbearing. Worldwatch Paper No. 16. Washington, D.C.: Worldwatch Institute.

Nicholaides III, J. J., D. A. Bandy, P. A. Sánchez, and C. Valverde S. 1982. Continuous cropping potential in the Amazon. Paper presented at the Conference on "Frontier Expansion in Amazonia," Center for Latin American Studies, University of Florida, Gainesville, February 8–11, 1982. Typescript.

Nietschmann, B. Q. 1971. The study of indigenous food production systems: mere subsistence or merrily subsisting? *Revista Geográfica* 74:83–99.

Nietschmann, B. Q. 1972. Hunting and fishing focus among the Miskito Indians, Eastern Nicaragua. *Human Ecology* 1:41–67.

Nietschmann, B. Q. 1974. *Between Land and Water.* New York: Seminar Press.

Nogueira-Neto, P. and J. C. de M. Carvalho. 1979. A programme of ecological stations for Brazil. *Environmental Conservation* 6(6):95–104.
North Carolina State University. 1974. Soil Science Department. *Agronomic-Economic Research on Tropical Soils. Annual Report for 1974.* Raleigh: North Carolina State University.
A *Notícia* (Manaus). July 22, 1983. "Amazonas ganha nova reserva ecológica." Section 1, p. 3.
Nye, P. H. and D. J. Greenland. 1960. *The Soil Under Shifting Cultivation.* Technical Communication No. 51. Harpenden, U.K.: Commonwealth Agricultural Bureaux of Soils.
Nye, P. H. and D. J. Greenland. 1964. Changes in the soil after clearing tropical forest. *Plant and Soil* 21:101–12.
Ochmen, K. H. and W. Paul. 1974. Population models. Vols. 1 and 2. In *Proceedings of the Seminar on the Regionalized Multi-Level World Model at the International Institute for Applied Systems Analysis (IIASA), Laxenburg, Austria, April 29–May 3, 1974.* Laxenburg, Austria: IIASA.
Odum, E. P. 1969. The strategy of ecosystem development. *Science* 164:262–70.
Odum, E. P. 1971. *Fundamentals of Ecology.* 3d ed. Philadelphia: Saunders.
Odum, H. T. 1971. *Environment, Power, and Society.* New York: Wiley-Interscience.
Odum, H. T. 1983. *Systems Ecology: An Introduction.* New York: Wiley.
Oldfield, M. L. 1981. Tropical deforestation and genetic resources conservation. *Studies in Third World Societies* 14:277–345.
Pádua, M. T. J. 1976. Documento do Brasil. In *Reunião Internacional sobre administração de unidades de conservação na região amazônica, Santarém, Pará, Brasil, November 8–14, 1976,* Instituto Interamericano de Ciências Agrícolas da O.E.A. (IICA-TRÓPICOS)/Instituto Brasileiro de Desenvolvimento Florestal (IBDF), pp. III-A-1–III-A-54. Informes de Conferencias Cursos y Reuniones No. 107. Belém: IICA-TRÓPICOS.
Pádua, M. T. J. n.d. (c.1979). Parques Nacionais e reservas equivalentes. Brasília: Instituto Brasileiro de Desenvolvimento Florestal (IBDF).
Pádua, M. T. J., A. Magnanini, and R. A. Mittermeir. 1974. Brazil's national parks. *Oryx* 7:452–64.
Pádua, M. T. J. and A. T. B. Quintão. 1982. Parks and biological reserves in the Brazilian Amazon. *Ambio* 11(5):309–14.
Páez, G. and S. Dutra. 1974. Algumas considerações sobre o delineamento de sistemas de produção. In *Reunião sobre Diretrizes de Pesquisa Agrí-*

cola para a Amazônia (Trópico Úmido), Brasília, May 6–10, 1974. Empresa Brasileira de Pesquisa Agropecuária (EMBRAPA), 1:4.2–4.22. Brasília: EMBRAPA.

Palmer, E. R. 1973. Gmelina arborea as a potential source of hardwood pulp. *Tropical Science* 15:243–60.

Pandolfo, C. 1978. *A Floresta Amazônica Brasileira—Enfoque Econômico-Ecológico*. Belém: Superintendência do Desenvolvimento da Amazônia (SUDAM).

Pandolfo. C. 1979. Florestas de rendimento. Paper prepared for the Simpósio sobre Amazônia e seu Uso Agrícola, 31ª. Reunião da Sociedade Brasileira para o Progresso da Ciência, Fortaleza, Ceará, July 11–18, 1979. Mimeo. (Abstract).

Patten, B. C. 1971. A primer for ecological modeling and simulation with analog and digital computers. In B. C. Patten, ed., *Systems Analysis and Simulation in Ecology*, 1:3–121. New York: Academic Press.

Pearl, R. and L. J. Reed. 1920. On the rate of growth of the population of the United States since 1790 and its mathematical representation. *Proceedings of the National Academy of Sciences* 6:275–88.

Pearl, R., L. J. Reed, and K. F. Kish. 1940. The logistic curve and the census count of 1940. *Science* 92:486–88.

Pendleton, R. L. 1956. The place of tropical soils in feeding the world. In *Smithsonian Report for 1955*, pp. 441–58. Washington, D.C.: Smithsonian Institution.

Penteado, A. R. 1967. *Problemas de Colonização e Uso da Terra na Região Bragantina do Estado do Pará*. Belém: Universidade Federal do Pará.

Pereira, F. B. and J. de S. Rodrigues. 1971. *Possibilidades Agro-Climaticas do Município de Altamira (Pará)*. Belém: Departamento de Assuntos Universitários, Escola de Agronomia da Amazônia, Boletim No. 1.

Peru, IVITA. (Instituto Veterinario de Investigación del Trópico y Altura). 1976. *Instituto Veterinario de Investigación del Trópico y Altura, Presentación al Sr. Ministro de Alimentación*. Lima: Universidad Nacional Mayor de San Marcos.

Pianka, E. R. 1966. Latitudinal gradients in species diversity: A review of the concepts. *American Naturalist* 100:33–56.

Pianka, E. R. 1974. *Evolutionary Ecology*. New York: Harper and Row.

Pielou, E. C. 1969. *An Introduction to Mathematical Ecology*. New York: Wiley-Interscience.

Pimentel, D., L. E. Hurd, A. C. Bellotti, M. J. Forster, I. N. Oka, O. D. Scholes, and R. J. Whitman. 1973. Food production and the energy crisis. *Science* 182:443.

Pinheiro, F. P., G. Bensabath, A. H. P. Andrade, Z. C. Lins, H. Fraika, A. T. Tang, R. Lainson, J. J. Shaw, and M. C. Azevedo. 1974a. Infectious diseases along Brazil's Trans-Amazon Highway: Surveillance and research. *Pan American Health Organization Bulletin* 8(2):111–22.

Pinheiro, F. P., G. Bensabath, D. Costa, Jr., O. M. Maroja, Z. C. Lins, and A. H. P. Andrade. 1974b. Haemorrhagic syndrome of Altamira. *The Lancet* 1(7859):639–42.

Pinho Filho, E. 1979. *Amazônia Entre Contrastes*. Belém: Mitograph Editora Ltda.

Pires, J. M. 1973. Tipos de Vegetação da Amazônia. Belém: Museu Paraense Emílio Goeldi, Publicação Avulsa No. 20, pp. 179–202.

Pires, J. M. 1978. The forest ecosystems of the Brazilian Amazon: description, functioning and research needs. In: *Tropical Forest Ecosystems: a State of Knowledge Report*. United Nations Educational Scientific and Cultural Programme/United Nations Environmental Programme/United Nations Food and Agriculture Organization (UNESCO/UNEP/FAO). pp. 607–27. Paris: UNESCO.

Pires, J. M. and G. T. Prance. 1977. The Amazon forest: A natural heritage to be preserved. In G. T. Prance and E. S. Elias, eds., *Extinction Is Forever*, pp. 158–94. Bronx: New York Botanical Garden.

Pool, D. J. 1972. *Insect Leaf Damage as Related to the Intensity of Management in Tropical Wet Forest Succession*. M.S. thesis in Agronomy, University of Florida, Gainesville.

Poore, D. 1976. The values of tropical moist forest ecosystems. *Unasylva* 28:128–45.

Popenoe, H. 1960. *Effects of Shifting Cultivation on Natural Soil Constituents in Central America*. Ph.D. dissertation, University of Florida, Gainesville.

Posey, J. W. and P. F. Clapp. 1964. Global distribution of normal surface albedo. *Geophysics International* 4:53–58.

Potter, G. L., H. W. Ellsaesser, M. C. McCracken, and F. M. Kuther. 1975. Possible climatic impact of tropical deforestation. *Nature* 258:697–98.

Prance, G. T. 1975. Flora and Vegetation. In R. J. A. Goodland and H. S. Irwin, *Amazon Jungle: Green Hell to Red Desert? An Ecological Discussion of the Environmental Impact of the Highway Construction Program in the Amazon Basin*, pp. 101–11. New York: Elsevier.

Prance, G. T., W. A. Rodrigues, and M. F. da Silva. 1976. Inventário florestal de um hectare de mata de terra firme km 30 Estrada Manaus-Itacoatiara. *Acta Amazonica* 6:9–35.

A Província do Pará (Belém). December 4, 1974. "Floresta Nacional do Tapajós ameaçada de ficar no papel."

Ramanathan, V. 1981. The role of ocean-atmosphere interaction in the CO_2 climate problem. *Journal of the Atmospheric Sciences* 38:918–30.

Ramos, A. R. 1980. Development, integration, and the ethnic integrity of Brazilian Indians. In F. Barbira-Scazzocchio, ed., *Land, People and Planning in Contemporary Amazonia*, Occasional Paper No. 3, pp. 222–29. Cambridge, U.K.: Cambridge University, Centre of Latin-American Studies.

Randers, J. and D. H. Meadows. 1972. The carrying capacity of the globe. *Sloan Management Review* 15(2):11–27.

Rankin, J. M. 1979. Manejo florestal ecológico. *Acta Amazonica* 9(4):115–22 (suplemento).

Rappaport, R. A. 1968. *Pigs for the Ancestors: Ritual in the Ecology of a New Guinea People*. New Haven: Yale University Press.

Rappaport, R. A. 1971. The flow of energy in an agricultural society. *Scientific American* 224:116–32.

Rebelo, D. C. 1973. *Transamazônica: Integração em Marcha*. Rio de Janeiro: Ministério de Transportes, Centro de Documentação e Publicações.

Reed, L. J. 1936. Population growth and forecasts. *Annals of the American Academy of Political and Social Science*, November 1936.

Reis, A. C. F. 1972. *A Amazônia e a Cobiça Internacional*. 4th ed. Rio de Janeiro: Companhia Editora Americana.

Reis, M. S. 1978. Uma definição técnico-política para o aproveitamento racional dos recursos florestais da Amazônia brasileira. Conferência proferida durante o 3°. Congresso Florestal Brasileiro, Manaus, Amazonas, December 4–7, 1978. Brasília: Projeto de Desenvolvimento e Pesquisa Florestal (PRODEPEF)/Instituto Brasileiro de Desenvolvimento Florestal (IBDF).

Revelle, R. 1982. Carbon dioxide and world climate. *Scientific American* 247(2):33–41.

Ribeiro, G. T. and R. A. Woessner. 1978. Teste da eficiencia com seis (6) sauvicidas no controle de sauvas *Atta* spp. na Jari. In *Procedimentos do V Congresso Brasileiro de Entomologia, Seção de Controle Químico*, Ilheus & Itabuna, Bahía, Brasil, July 1978.

Richards, P. W. 1964. *The Tropical Rain Forest*. 2d ed. Cambridge: Cambridge University Press.

Ridker, R. G. 1976. Perspectives on population policy and research. In R. G. Ridker, ed., *Population and Development: The Search for Selective*

Interventions, pp. 1–35. Baltimore: Johns Hopkins University Press.
Roff, D. A. 1974. Spatial heterogeneity and the persistence of populations. *Oecologia* 15:245–58.
Rosenn, K. S. 1971. The jeito: Brazil's institutional bypass of the formal legal system and its development implications. *American Journal of Comparative Law* 19:514–49.
Ross, M. A. 1980. The role of land clearing in Indonesia's transmigration program. *Bulletin of Indonesian Economic Studies* 16(1):75–85.
Ruderman, M. A., H. M. Foley, and J. W. Chamberlain. 1976. Eleven-year variation in polar ozone and stratospheric-ion chemistry. *Science* 192:555–57.
Rundel, T. K. 1983. Roads, speculators, and colonization in the Ecuadorian Amazon. *Human Ecology* 11(4):385–403.
Russell, E. W. 1973. *Soil Conditions and Plant Growth*. 10th ed. London: Longman.
Ruthenberg, H. 1971. *Farming Systems in the Tropics*. Oxford, U.K.: Clarendon Press.
Rylands, A. B. 1984. Primate conservation areas in Brazilian Amazonia. *Primate Eye* 24:22–25.
Rylands, A. B. and R. A. Mittermeir. 1982. Conservation of primates in Brazilian Amazonia. *International Zoo Yearbook* 22:17–37.
Sagan, C., O. B. Toon, and J. B. Pollack. 1979. Anthropogenic albedo changes and the earth's climate. *Science* 206:1363–68.
Sahlins, M. D. 1972. *Stone Age Economics*. Chicago: Aldine-Atherson.
Salati, E., J. Marques, and L. C. B. Molion. 1978. Origem e distribuição das chuvas na Amazônia. *Interciencia* 3:200–6.
Salati, E. and P. B. Vose. 1984. Amazon Basin: a system in equilibrium. *Science* 225:129–38.
Sánchez, P. A. 1973. Soil management under shifting cultivation. In P. A. Sánchez, ed., *A Review of Soils Research in Tropical Latin America*, pp. 62–92. Raleigh: North Carolina State University, Soil Science Department.
Sánchez, P. A. 1976. *Properties and Management of Soils in the Tropics*. New York: Wiley.
Sánchez, P. A. 1977. Advances in the management of OXISOLS and ULTISOLS in tropical South America. In *Proceedings of the International Seminar on Soil Environment and Fertility Management in Intensive Agriculture, Tokyo, Japan*, pp. 535–66. Tokyo: Society of the Science of Soil and Manure.
Sánchez, P. A., D. E. Bandy, J. H. Villachica, and J. J. Nicholaides, III.

1982. Amazon basin soils: Management for continuous crop production. *Science* 216:821-27.

Sánchez, P. A. and S. W. Buol. 1975. Soils of the tropics and the world food crisis. *Science* 188:598-603.

Sanders, T. G. 1971. Brazilian interior migration. *American Field Staff Reports, East Coast South America Series* 15(2):1-10.

Sanders, T. G. 1973. Colonization on the Transamazonian Highway. *American Field Staff Reports, East Coast South America Series* 17(3):1-9.

Saunders, J. 1974. The population of the Brazilian Amazon today. In C. Wagley, ed., *Man in the Amazon*, pp. 160-80. Gainesville: University Presses of Florida.

Schacht, R. M. 1980. Two models of population growth. *American Anthropologist* 82(4):782-98.

Schmink, M. 1982. Land conflicts in Amazonia. *American Ethnologist* 9(2):341-57.

Schmithüsen, F. 1978. *Contratos de Utilização Florestal com Referência Especial a Amazônia Brasileira.* (PNUD/FAO/IBDF/BRA/76/027 Série Técnica No. 12). Brasília: Projeto de Desenvolvimento e Pesquisa Florestal (PRODEPEF).

Schubart, H. O. R., W. J. Junk, and M. Petrere Jr. 1976. Sumário de ecologia Amazônica. *Ciência e Cultura* 28(5):507-9.

Schware, R. and W. W. Kellogg. 1982. De como pudieran influir los cambios de clima sobre la producción de alimentos. *Ceres* 15(2):40-42.

Scott, G. A. J. 1975. Soil profile changes resulting from the conversion of forest to grassland in the Montaña of Peru. *Great Plains-Rocky Mountain Geographical Journal* 4:124-30.

Scott, G. A. J. 1978. *Grassland Development in the Gran Pajonal of Eastern Peru: a Study of Soil-Vegetation Nutrient Systems.* Hawaii Monographs in Geography, No. 1. Honolulu, University of Hawaii at Manoa, Department of Geography.

Seiler, W. and P. J. Crutzen. 1980. Estimates of gross and net fluxes of carbon between the biosphere and the atmosphere from biomass burning. *Climatic Change* 2:207-47.

Serete S. A. Engenharia and Brazil Ministério do Interior, Superintendência do Desenvolvimento da Amazônia (SUDAM). 1972. *Estudos Setoriais e Levantamento de Dados da Amazônia.* Vol. II-3 (Setores Econômicos e Elementos da Política Setorial: Pecuária Bovina), pp. 11-23. Belém: SUDAM.

Serrão, E. A. S. 1979. Produtividade das pastagens cultivadas na região Amazônica. Paper presented at the Simpósio sobre Amazônia e o seu uso Agrícola, 31a. Reunião Anual da Sociedade Brasileira para o Progresso da Ciência, Fortaleza, Ceará, July 16–17, 1979. Mimeo. (Abstract).

Serrão, E. A. S., E. de S. Cruz, M. Simão Neto, G. F. de Sousa, J. B. Bastos, and M. C. de F. Guimarães. 1971. Resposta de três gramíneas forrageiras (*Brachiaria decumbens* Stapf., *Brachiaria ruziziensis* Germain et Everard e *Pennisetum purpureum* Schum.) a elementos fertilizantes em latossolo amarelo textura media. *Instituto de Pesquisa Agropecuária do Norte (IPEAN), Série: Fertilidade do Solo* 1(2):1–38.

Serrão, E. A. S. and I. C. Falesi. 1977. *Pastagens do Trópico Úmido Brasileiro*. Belém: Empresa Brasileira de Pesquisa Agropecuária–Centro de Pesquisa Agropecuária do Trópico Úmido (EMBRAPA-CPATU).

Serrão, E. A. S., I. C. Falesi, J. B. da Viega, and J. F. Teixeira Neto. 1978. *Produtividade de Pastagens Cultivadas em Solos de Baixa Fertilidade das Áreas de Floresta do Trópico Úmido Brasileiro*. Belém: Empresa Brasileira de Pesquisa Agropecuária–Centro de Pesquisa Agropecuária do Trópico Úmido (EMBRAPA-CPATU).

Serrão, E. A. S., I. C. Falesi, J. B. de Viega, and J. F. Teixeira Neto. 1979. Productivity of cultivated pastures on low fertility soils in the Amazon of Brazil. In P. A. Sánchez and L. E. Tergas, eds., *Pasture Production in Acid Soils of the Tropics: Proceedings of a Seminar held at CIAT, Cali, Colombia, April 17–21, 1978*, pp. 195–225. Cali, Colombia: Centro Internacional de Agricultura Tropical (CIAT). Series 03 EG-05.

Shantzis, S. B. and W. W. Behrens, III. 1973. Population control mechanisms in a primitive agricultural society. In D. L. Meadows and D. H. Meadows, eds., *Toward Global Equilibrium, Collected Papers*, pp. 257–88. Cambridge, Mass.: Wright-Allen Press.

Simão Neto, M., E. A. S., Serrão, C. A. Gonçalves, and D. M. Pimentel. 1973. *Comportamento de Gramíneas Forrageiras na Região de Belém*. Belém: Instituto de Pesquisas Agropecuárias do Norte (IPEAN) Comunicado Técnico No. 44.

Simon, J. L. 1976. Income, wealth, and their distribution as policy tools in fertility control. In R. G. Ridker, ed., *Population and Development: The Search for Selective Interventions*, pp. 36–76. Baltimore: Johns Hopkins University Press.

Simpson, G. J. 1975. A *Report on the Karimui Resettlement Scheme: Prob-*

lems and Prospects. Port Moresby, Papua New Guinea: Office of Environment and Conservation.

Sioli, H. 1973. Recent human activities in the Brazilian Amazon Region and their ecological effects. In B. J. Meggers, E. S. Ayensu, and W. D. Duckworth, eds., *Tropical Forest Ecosystems in Africa and South America: A Comparative Review.* pp. 321–34. Washington, D.C.: Smithsonian Institution Press.

Sioli, H. 1980. Foreseeable consequences of actual development schemes and alternative ideas. In F. Barbira-Scazzocchio, ed., *Land, People, and Planning in Contemporary Amazonia.* Occasional Paper No. 3, pp. 252–68. Cambridge, U.K.: Cambridge University, Centre of Latin-American Studies.

Skillings, R. F. and N. O. Tcheyan. 1979. Economic development prospects of the Amazon Region of Brazil. Baltimore: Center of Brazilian Studies, School of Advanced International Studies, Johns Hopkins University, Occasional Paper No. 9.

Smith, N. J. H. 1976a. Brazil's Transamazon Highway settlement scheme: Agrovilas, agropoli, and ruropoli. *Association of American Geographers Proceedings* 8:129–32.

Smith, N. J. H. 1976b. *Transamazon Highway: A Cultural-Ecological Analysis of Colonization in the Humid Tropics.* Ph.D. dissertation, University of California, Berkeley.

Smith, N. J. H. 1978. Agricultural productivity along Brazil's Transamazon Highway. *Agro-Ecosystems* 4:415–32.

Smith, N. J. H. 1981. Colonization lessons from a rainforest. *Science* 214:755–61.

Smith, N. J. H. 1982. *Rainforest Corridors: the Transamazon Colonization Scheme.* Berkeley: University of California Press.

Snedecor, G. W. and W. G. Cochran. 1967. *Statistical Methods.* 6th ed. Ames: Iowa State University Press.

Schneider, S. H., W. W. Kellogg, and V. Ramanathan. 1980. Carbon dioxide and climate. *Science* 210:6–7.

Sombroek, W. G. 1966. *Amazon Soils: A Reconnaisance of the Soils of the Brazilian Amazon Region.* Wageningen, Holland: Centre for Agricultural Publications and Documentation.

Sommer, A. 1976. Attempt at an assessment of the world's tropical moist forests. *Unasylva* 28(112–13):5–24.

Stark, N. 1970. Direct nutrient cycling in the Amazon Basin. In J. M. Idrobo, ed., *II Simpósio de Biologia Tropical Amazonica,* pp. 172–77. Bogotá: Asociación de Biologia Tropical.

Stark, N. 1971. Mycorrhizae and nutrient cycling in the tropics. In E. Hacskaylo, ed., *Mycorrhizae: Proceedings of the First North American Conference on Mycorrhizae, April 1969*, pp. 228–29. United States Department of Agriculture Miscellaneous Publication No. 1189. Washington, D.C.: GPO.

Stark, N. 1972. Nutrient cycling pathways in litter fungi. *BioScience* 22:355–60.

Stout, P. R. 1974. Agriculture's energy requirements. In D. E. McCloud, ed., *A New Look at Energy Sources*, pp. 13–22. American Society of Agronomy Special Publication No. 22. Madison, Wisc.: American Society of Agronomy.

Street, J. M. 1969. An evaluation of the concept of carrying capacity. *Professional Geographer* 21:104–7.

Struchtemeyer, R. A., D. M. Chaves, G. F. de Sousa, E. de S. Cruz, and J. C. A. J. de Magalhães. 1971. Necessidade de calcário em solos da zona Bragantina. *Instituto de Pesquisas e Experimentação Agropecuárias do Norte (IPEAN) Série: Fertilidade do Solo* 1(1):1–21.

Stuiver, M. 1978. Atmospheric carbon dioxide and carbon reservoir changes. *Science* 199:253–58.

Tambs, L. A. 1974. Geopolitics of the Amazon. In C. Wagley, ed., *Man in the Amazon*, pp. 45–87. Gainesville: University of Florida Presses.

Tamer, A. 1970. *Transamazônica, Solução para 2001*. Rio de Janeiro: APEC Editôra.

Tardin, A. T., A. P. dos Santos, D. C. L. Lee, F. C. S. Maia, F. J. Mendonça, C. V. Assunção, J. E. Rodrigues, M. de Moura Abdon, R. A. Novaes, S. C. Chen, V. Duarte, and Y. E. Shimabukuro. 1979. *Levantamento de áreas de Desmatamento na Amazônia Legal Através de Imagens de Satélite LANDSAT*. (INPE-COM3/NTE,C.D.U. 621.38SR). São José dos Campos. São Paulo: Instituto Nacional de Pesquisas Espaciais (INPE).

Tardin, A. T., A. P. dos Santos, E. M. I. Moraes Novo, and F. L. Toledo. 1978. Projetos agropecuários da Amazônia; desmatamento e fiscalização—relatório: A *Amazônia Brasileira em Foco* 12:7–45.

Tardin, A. T., D. C. L. Lee, R. J. R. Santos, O. R. de Assis, M. P. dos Santos Barbosa, M. de Lourdes Moreira, M. T. Pereira, D. Silva, and C. P. dos Santos Filho. 1980. *Subprojeto Desmatamento, Convênio IBDF/CNPq-INPE 1979*. Instituto Nacional de Pesquisas Espaciais (INPE) Relatório No. INPE-1649-RPE/103. São José dos Campos, São Paulo: INPE.

Tavares, V. P., C. M. C. Considera, and M. T. I. I. de C. Silva. 1972.

Colonização Dirigida no Brasil, suas Possibilidades na Região Amazônica. Instituto de Planejamento Econômico e Social (IPEA)/Instituto de Pesquisas (INPES) Relatório de Pesquisa No. 8. Rio de Janeiro: IPEA/INPES.

Teitelbaum, M. S. 1975. Relevance of demographic transition theory for developing countries. *Science* 188:420-25.

Thery, H. 1976. *Rondônia: Mutations d'un Territoire Fédéral en Amazonie Brésilienne.* Thèse de 3e cycle. Paris: Université de Paris I, C. N.R.S. 111, Ecole Normale Supérieure.

Thomas, R. H., T. J. O. Sanderson, and K. E. Rose. 1979. Effect of climatic warming on the West Antarctic ice sheet. *Nature* 277:355-58.

Toledo, J. M. and E. A. S. Serrão. 1982. Pasture and animal production in Amazonia. In, S. B. Hecht, ed., *Amazonia: Land Use and Agricultural Research.* pp. 281-309. Cali, Colombia: Centro Internacional de Agricultura Tropical (CIAT).

UNEP (United Nations Environmental Programme). 1980. *Overview Document, Experts Meeting on Tropical Forests, Nairobi, 25 Feb.-1 March 1980.* UNEP/WG.34/4.

UNESCO (United Nations Educational, Scientific, and Cultural Organization). 1977. UNESCO/United Nations Fund for Population Activities (UNFPA). *Population, Resources, and Development in the Eastern Islands of Fiji: Information for Decision-Making.* General Report No. 1 of UNESCO/UNFPA Population and Environment Project in the Eastern Islands of Fiji. Man and the Biosphere (MAB) Programme, Project 7: Ecology and Rational Use of Island Ecosystems. Paris: UNESCO.

UNESCO (United Nations Educational, Scientific, and Cultural Organization). 1978. UNESCO/United Nations Environmental Programme (UNEP)/United Nations Food and Agricultural Organization (UN-FAO). *Tropical Forest Ecosystems: A State of Knowledge Report.* Paris: UNESCO.

UN-FAO (United Nations Food and Agriculture Organization). 1959. Shifting cultivation—FAO's position and course of action. *Proceedings of the Ninth Pacific Science Congress* 7:71.

UN-WHO (United Nations World Health Organization). 1973. *Energy and Protein Requirements.* World Health Organization Technical Report Series. No. 552. Geneva: UN-WHO.

U.S. CEQ/DS (U.S. Council on Environmental Quality and Department of State). 1980. *The Global 2000 Report to the President.* 3 vols. New York: Pergamon Press.

U.S. DAgr. (United States Department of Agriculture). 1960. Soil Survey Staff. *Soil Classification; A Comprehensive System—7th Approximation.* Washington, D.C.: GPO.

U.S. DAgr. (United States Department of Agriculture). 1965. *Changes in Agriculture in 26 Developing Nations, 1948 to 1963.* Foreign Agricultural Report No. 27. Washington, D.C.: GPO.

U.S. DS (United States Department of State). 1978. *Proceedings of 1978 U.S. Strategy Conference on Tropical Deforestation.* Washington, D.C.: GPO.

U.S. NAS (United States, National Academy of Sciences). 1979. *Carbon Dioxide and Climate: A Scientific Assessment.* Washington, D.C.: NAS Press.

U.S. NAS (United States, National Academy of Sciences). 1982a. *Carbon Dioxide and Climate: A Second Assessment.* Washington, D.C.: NAS Press.

U.S. NAS (United States, National Academy of Sciences). 1982b. *Causes and Effects of Stratospheric Ozone Reduction: An Update.* Washington, D.C.: NAS Press.

Valverde, O. 1979a. Considerações finais e conclusões. In O. Valverde, ed., *A Organização do Espaço na Faixa da Transamazônica,* vol. 1: *Introdução, Sudoeste Amazônico e Regiões Vizinhas,* pp. 245–58.

Valverde, O., ed. 1979b. *A Organização do Espaço na Faixa da Transamazônica,* vol. 1: *Introdução Sudoeste amazônico, Rondônia e Regiões Visinhas.* Rio de Janeiro: Instituto Brasileiro de Geografia e Estatística (IBGE).

Valverde, O., and C. V. Dias. 1967. *A Rodovia Belém- Brasília: Estudo de Geografia Regional.* Rio de Janeiro: Instituto Brasileiro de Geografia e Estatística (IBGE).

Valverde S., C. and D. E. Bandy. 1982. Production of annual food crops in the Amazon. In S. B. Hecht, ed., *Amazonia: Agriculture and Land-Use Research,* pp. 243–80. Cali, Colombia: Centro Internacional de Agricultural Tropical (CIAT).

Van Valen, L. 1971. The history and stability of atmospheric oxygen. *Science* 171:439–43.

Van Velthem, L. H. 1980. O Parque indigina de Tumucumaque. *Boletim do Museu Paraense Emílio Goeldi,* no. 76, pp. 1–31.

Van Wambeke, A. 1978. Properties and potentials of soils in the Amazon Basin. *Interciencia* 3:233–41.

Vayda, A. P. 1969. An ecological approach in cultural anthropology. *Bucknell Review* 17(1):112–19.

Vayda, A. P. 1976. On the "New Ecology" paradigm. *American Anthropologist* 78:645–46.

Vayda, A. P. and B. J. McCay. 1975. New directions in ecology and ecological anthropology. *Annual Review of Anthropology* 4:293–306.

Veja (Rio de Janeiro). 1980. "Censo: conta baixa. O IBGE revê o exagero de suas previsões," November 12, p. 28.

Veja (Rio de Janeiro). 1981. "Os Arara saem da mata." March 11, pp. 72–76.

Veja (Rio de Janeiro). 1982. "Corte no verde: uma estrada rasgará o Parque do Araguaia," December 22, p. 90.

Verdade, F. da C. 1974. Problemas de fertilidade do solo na Amazônia. *Ciência e Cultura* 26:219–24.

Verhulst, P. F. 1838. Notice sur la loi que la population suit dans son accroissement. *Correspondence Mathematique et Physique* 10:113–21. Abridged translation in E. Kormondy, ed., *Readings in Ecology*, pp. 64–66. Englewood Cliffs, N.J.: Prentice Hall, 1965.

Vermeer, D. E. 1970. Population pressure and crop rotational changes among the Tiv of Nigeria. *Annals of the Association of American Geographers* 60:299–314.

Vicente-Chandler, J. 1975. Intensive management of pastures and forages in Puerto Rico. In E. Bornemisza and A. Alvarado, eds., *Soil Management in Tropical America: Proceedings of a Seminar held at CIAT, Cali, Colombia, February 10–14, 1974*, pp. 409–52. Raleigh: North Carolina State University, Soil Science Department.

Viégas, R. M. F. and D. C. L. Kass. 1974. *Resultados de Trabalhos Experimentais na Transamazônica no Período de 1971 a 1974*. Belém: Empresa Brasileira de Pesquisa Agropecuária, Instituto de Pesquisa Agropecuária do Norte (EMBRAPA-IPEAN).

Villa Nova, N. A., E. Salati, and E. Matusi. 1976. Estimativa da evapotranspiração na Bacia Amazônica. *Acta Amazônica* 6(2):215–28.

Waddell, E. 1972. *The Mound Builders: Agricultural Practices, Environment, and Society in the Central Highlands of New Guinea*. American Ethnological Society Monograph No. 53. Seattle: University of Washington Press.

Wade, N. 1979. CO_2 in climate: gloomsday predictions have no fault. *Science* 206:912–13.

Wagley, C. 1976 (1953). *Amazon Town: A Study of Man in the Tropics*, 2d ed. New York: Macmillan.

Watt, K. E. F. 1966. *Systems Analysis in Ecology*. New York: Academic Press.

Watters, R. F. 1971. *Shifting Cultivation in Latin America*. Food and Agriculture Organization of the United Nations (UN-FAO) Forestry Development Paper No. 17. Rome: UN-FAO.

Waugh, D. L., R. B., Cate, L. A. Nelson, and A. Manzano. 1975. New concepts in biological and economical interpretation of fertilizer response. In E. Bornemisza and A. Alvarado, eds., *Soil Management in Tropical America: Proceedings of a Seminar held at CIAT, Cali, Colombia. February 10-14, 1974*, pp. 484-563. Raleigh: North Carolina State University, Soil Science Department.

Weisman, T. 1974. A model for the relationship between selected nutritional variables and excess mortality in populations. In *Proceedings of the Seminar on the Regionalized Multi-Level World Model at the International Institute for Applied Systems Analysis (IIASA), Laxenburg, Austria. April 29-May 3, 1974*, pp. B481-98. Laxenburg, Austria: IIASA.

Weiss, R. F. 1981. The temporal and spatial distribution of tropospheric nitrous oxide. *Journal of Geophysical Research* 86:7185-95.

Weiss, R. F. and H. Craig. 1976. Production of atmospheric nitrous oxide from combustion. *Geophysical Research Letters* 3:751-53.

Went, F. W. and N. Stark. 1968. The biological and mechanical role of soil fungi. *Proceedings of the National Academy of Sciences* 173:171-80.

Wesche, R. 1974. Planned rainforest family farming on Brazil's Transamazonic Highway. *Revista Geografica* 81:105-14.

Whittaker, R. H. and G. E. Likens. 1973. primary production: The biosphere and man. *Human Ecology* 1:357-69.

Wilbur, H. M. 1972. Competition, predation, and the structure of the *Ambystoma, Rana sylvatica* community. *Ecology* 53:3-21.

Williams, G. C. 1966. *Adaptation and Natural Selection*. Princeton, N.J.: Princeton University Press.

Wilson, E. O. and W. H. Bossert. 1971. A *Primer of Population Biology*. Stamford, Conn.: Sinaur.

Wood. C. H. and M. Schmink. 1979. Blaming the victim: small farmer production in an Amazon colonization project. *Studies in Third World Societies* 7:77-93.

Woodwell, G. M. 1978. The carbon dioxide question. *Scientific American* 238:34-43.

Woodwell, G. M., J. E. Hobbie, R. A. Houghton, J. M. Melillo, B. Moore, B. J., Peterson, and G. R. Schaver. 1983. Global deforestation: contribution to atmospheric carbon dioxide. *Science* 222:1081-86.

Woodwell, G. M., R. H. Whittaker, W. A. Reiners, G. E. Likens, C. C. Delwiche, and D. P. Botkin. 1978. The biota and the world carbon budget. *Science* 199:141–46.

Index

Abandonment: of fields in shifting cultivation, 54; of lots, 25, 117
Aboriginal populations: 84-85; *see also* Amerindians
ACAR-PARÁ, 114, 231
Acculturation, 140
Acidity, *see* pH
Acre: colonization in, 19; deforestation in, 4-5; rubber germplasm in, 51
Affluence, 155
Africa, 83, 98
Age: composition, 132; effect on labor, 116; of lot owners, 218; *see also* demography
Agrarian reform, 153
Agregado, 97, 231
Agricultural experience, 117
Agricultural extension, 100
Agricultural intensity, 81
Agricultural potential, 155
Agricultural production sector, 128
Agricultural year, 25, 231
Agriculture: effect of credit bureaucracy on, 168; operations in, 176–84
AGRISIM, 125, 130, 231
Agroecosystem, 61, 62, 231; installation cost of, 148
Agrópolis, 20, 97, 231
Agrovila, 20, 93, 231
Albedo, 49-51
Alcohol, 153
ALFISOL, 38-39, 51, 94, 97
Allan, W., 15, 73-74, 81, 228nn3.2,3.5
Allee effect, 79
Allocation: to crops, 132
Altamira: city of, 19, 97, 117, 157; colonization area, 39, 93, 20; rainfall in, 99
Altitude: effect on production, 98-99
Aluminium: effect on rice yields, 106; change from agriculture, 59; change from liming, 198; change from second growth burns, 190; change from virgin burns, 189; change from weed burns, 191; change under pasture, 195; change without burning, 192; in initial soil, 161, in soil under pasture, 55
Alvim, P. de T., 31
Amapá, 28, 36; deforestation in, 5
Amazonas (State), 36; deforestation in, 5
Amazonia, 231
Amazonia National Park, 9
Amerindians, 231; and anthropogenic soil, 38; disappearance of, 52-53; land preparation practices of, 227n8; presence in area, 229n4.3; reserves for, 7-9, 149; *see also* Aboriginal peoples
Analytical models, 86
Animal protein, 82, 114-15
Animals: barnyard, 21, 132
Annual cash crops: land use strategy, 111-12
Annual crops, 24, 103; effects on soil, 53, 59; erosion under, 199; continuous cultivation of, 58; technology change in, 109
Ants: leafcutter, 32; *see also* Atta
Araguaia, 8
Arara Indians, 22
Archaeologists: approach to carrying capacity, 228n3.1
Archaeology, 73
Aripuanã, 8
Armour-Swift, 27
Artisan farmers, 111
Aschmann, H., 84
Asia, 26
Aswan Dam, 156
Atta, 32; *see also* Ants, leaf cutter

Backgrounds: of colonists, 132
Banco do Brasil, 216, 231; *see also* Credit; Financing
Bank: transportation cost to, 212
Barbalha, 105, 118
Bare soil: defined, 171; erosion in, 199
BASA (Banco da Amazonia, SA), 28, 231
Bayliss-smith, T., 15, 78, 83

Beans, 41; allocation to, 141; definition of, 237; cash requirements for planting, 183; expected yield for, 172; financing of, 165-67; in colonist diet, 115; labor requirements for, 177-78, 180; price of, 213; seed requirement, 217; subsistence need for, 214; choice type of, 173; yield of, 102, 201-4
Behavior mode, 231
Belém, 4, 57
Belém-Brasília Highway, 16, 23, 26, 55, 57, 232; deforestation on, 4, 227n3
Belterra, 19
Bertholetia exelsa, 18; *see also* Brazil nuts
Bethlehem Steel, 31
Biases: cultural, 85
Biological control, 62
Biological reserves, 11; *see also* Reserves
Biomass: and global carbon problem, 44-45
Birth rate: *see* Fertility
Births, 132
Bitter manioc: as cash crop, 112; *see also* Manioc
Black flies, 118; *see also* Pium, 208-9
Black pepper, 19, 24; allocation to, 141; as cash crop, 112; cash requirements for planting, 184; disease in, 107, 208-9; erosion under, 200; extent of, 30; fertilizing of, 123; fertilizer and lime for, 195-96; financing of, 165-67; labor requirements for, 177-78, 181-82; price of, 213; technology change in, 109; yields of, 102, 107-9, 123, 132
Blackpod disease, 109
Black water rivers, 35
Bolivia, 7
Boulding, K., 17
Brachiaria decumbens: production of, 205-6
Brasília, 16
Brazil nuts, 18
Breeding; of crops, 107
BR-80 Highway, 8
Broca, 23; *see also* Underclearing
Brown, H., 17
Brush, S., 228n3.2
Buffers: against failure, 81, 114, 125, 132
Bulk density, 54; *see also* Compaction
Bureaucracy, 98, 125; effect on agriculture, 168
Burning, 26, 61, 102; and carbon dioxide, 44-45; and cultural tradition, 48; and nitrous oxide, 47; dates of, 122, 186; effects on soil, 53, 56, 59, 98, 123, 132, 188-92; period rainfalls, 164
Burn quality, 79, 103, 122-23, 132, 139; effect on land use choice of Amerindians, 227n8; prediction for second growth burns, 187-88; prediction for virgin burns, 186-87

Caboclo, 10, 156, 232
Cabo Orange National Park, 9
Cacao, 24; allocation to, 141-42; as cash crop, 112; cash requirements for, 183-84; disease in, 107; erosion under, 200; extent of, 30; fertilizer and lime for, 195, 197, 123; financing of, 165-67; labor requirements for, 177-78, 181-82; modeling yields of, 107, 109-10; price of, 213; productive life of, 109; technology change in, 109; yield of, 102, 123, 132
Calcium: effects of agriculture on, 59; effects of burning on, 53; in soil under pasture, 55; analysis method, 102
Calories, 82-83, 130, 133-34, 136; and population density, 137; effect on mortality, 116; consumed by colonists, 114-15; requirements by age class, 225
Campina, 36, 232
Campinarana, 9, 36, 232
Canela de ferro, 104-5
Capina, 232; *see also* Weeding
Capital, 113; allocation of, 124; and colonist turnover, 119; constraints on allocations, 25, 125, 176, 183-85; initial, 132, 222; investment of, 217; opportunity cost of, 151; sufficiency checks, 132
Capital goods: return on relative to manual labor, 219; use destination of, 218
Capoeira, 25, 232; *see also* Second growth
Carbon (in soil): changes from virgin burns, 190; changes under pasture, 194; changes without burning, 193-94; effects of agriculture on, 59; effect on rice yields, 106; in initial soil, 163; analysis method, 102
Carbon Dioxide: and deforestation, 43-44; greenhouse effect, 228n2.2
Carneiro, R., 75, 80, 82, 84-85, 228nn3.2, 228n3.4
Carrying capacity: and development policy, 155-57, and planning, 153-54; definitions

INDEX   283

of, 232; dynamic, 70; effect of variability on, 88; for shifting cultivation, 81; logistic, 65, 71, 73; instantaneous, 70-73, 77, 84; of pastures, (see Pastures, feeding capacity); operational definition of, 79; perceived, 78; short term, 18; static, 70; sustainable, 70, 73-80, 84, 121
Casa da farinha, 232
Cash: for food purchase, 114; requirements, 113; subsistence need for, 214
Cash crops, 110, 148; allocation decisions for, 185-86; probabilities of land use for, 112
Cash economy, 100
Cash standard of living, 130, 135-36; and population density, 137
Cassava, *see* Manioc
Cation exchange capacity, 232; effects of agriculture on, 59
Cattle, 26-28; financing of, 165-67; slaughter weight of, 206, yield of, 205-7; *see also* Pasture; Ranching
Causal loop diagrams; 128-29
Cecropia, 42
Central America, 98
CEPA, 232
CEPLAC (Executive Commission of the Cacao Growing Plan), 8, 232
Ceratocystis fimbriata, 32
Cerrado, 8-13, 28, 50, 154, 232
Charcoal: and global carbon problem, 45
Chemical inputs; costs of, 183-84
Chickens, 124, 214, 216
Children, outside labor of, 210
Chimbu Province, 15
CIBRAZEM (Brazilian Storage Company), 212, 232
Cities: emigration to, 226; *see* Urban centers
Class: social and economic, 98
Clay: in initial soil, 162
Clearing, 132; cash requirements for, 183; classes, 171-72; decisions, 170-72; effects on soil nutrients, 98; labor requirements for, 177-79; probabilities of, 140; proportion of land, 130; simulated, 135, 138
Climate: and deforestation, 43-51; stability of, 149; of study area, 95
Climax, 61
Clothing: cash need for, 214
Club of Rome, 116, 233
Coevolved relationships, 7, 52, 149, 233

Coffea arabica, 51; *see also* Coffee
Coffee, 31, 153; killed by frost, 118; price and consumption of, 215
Coivara, 26, 165, 227n8, 233
Colocasia spp., 79; *see also* Taro
Colombia, 7
Colonião, 27; *see also* Guinea grass
Colonist failure, *see* Failure
Colonist population: variability in, 101
Colonists: backgrounds of, 26, 100; selection of, 153; turnover in, 117-19
Colonist types, 124, 133, 141-44; effect on family unit emigration, 226; effect on investment patterns, 217; effect on initial capital and capital goods, 222; effect on land use strategy, 111; effect on purchase of capital goods, 217; frequencies in population, 116
Colonization: history of, 18-19; motivations for, 7, 15-16; objectives of, 153; population absorption by, 157
Common property dilemma, 150
Compaction, 54, 119, 147; effects of agriculture on, 9; under pasture, 56-57
Competition, 69
Complexity, 90
Components, 89
Conklin, H. C., 74, 80, 228n3.2
Consumption, 71, 100; assumptions concerning, 72, 81; criteria for failure, 137; minimum levels, 121; of food, 114-15; standards for, 79
Continuous cultivation, 54, 58, 81
Cooperation, 79
Cooperatives, 22
Corn, *see* Maize
Corporations, 21, 26-27, 149, 156; logging, 33; *see also* Enterprises
Costa Rica: rainforest protection in, 7
COTRIJUI, 22
Cowpeas: expected yield for, 172; price of, 213; probability of planting, 173; seed requirement for, 217; yield of, 102, 203-04
CPATU-1 (Center for Research in Agriculture and Cattle Ranching in the Humid Tropics), 8, 102, 233
Credit, 16, 125, 165-70; *see also* Financing
Crinipellis perniciosa, 109-10; *see also* Witches' broom
Criteria: for carrying capacity, 84; multiple, 139

Critical population size, 74
Crop varieties, 51
Crop yields, 100, 128; modeling of, 200-9; pH dependence of, 139; with fertilizers, 109
Cruzeiro, 233
Cuiabá-Porto Velho Highway, 233; deforestation on, 6
Cuiabá-Santarém Highway, 233
Cultivation factor, 74
Cultivation period: maximum, 171
Cultural evolution, 77
Custeio: definition of, 233; financing of, 165-69

Death: of colonists, 118, 132; probability of by age class, 220-22; rate, 69; *see* Mortality
Debt: and colonist turnover, 118; repayment of, 113-14, 216-17
Decision-making: economic, 151
Deforestation: and extinctions, 50-52; and ranching, 27-28; concentration of, 4-6; environmental impacts of, 42-52, 149; and land tenure, 156; irreversibility of, 7; rate of, 3; simulated, 135; *see also* Clearing
Degradation: of resource base, 73; *see* environmental degradation; Soils, degradation of
Demarcation: of reserves, 155
Demographic transition, 66-68, 71
Demography, 100, 116, 220-21; *see also* Emigration; Immigration; Marriage; Mortality; Turnover of colonists
Dendê, 31; *see also* oil palm
DENPASA (Dende do Pará, SA), 31, 233
Density: dependent controls, 65-70; dependent effects, 64, 81; dependent response, 62; of human population, 63; independent controls, 68-70; independent factors, 81
Dependents: age distribution, 220; numbers, ages and sexes of, 218
Depreciation: of capital goods, 218-19
Desertification, 48-51
Deterministic models, 234
Deterministic runs, 126, 136-37, 145
Development: criteria for, 147-50; policy, 155-57; redirection of, 133
Diagrams: causal loop, 128-30; FORTRAN, 131
Diet, 133; of colonists, 114-15; quality of, 85
Dioscorea, 216

Discounting, 148
Discount rate, 150-51
Diseases (of crops), 54, 98, 123, 132, 147; in black pepper, 107, 108-9, 208-9; in cacao, 107, 109-10; in maize, 108, 201; in *Phaseolus*, 107, 202; in rice, 104-5; in *Vigna*, 203; organisms, 69
Diseases (of humans): probabilities by age and sex in colonist population, 223; respiratory, 224; *see also* Health
Diversity: genetic, 50-52, 149; of crops; 125; of species, 62, 149
Documentation: of KPROG2, 121
Dothidella ulei, 40; *see* Microcyclus ulei
Doubling time, 65
Drought, 16, 18, 99
Dry season, 41, 50, 95, 125; *see also* Burning, period rainfalls
Dunes, 37
Durable goods, 111, 124

Ecological Stations, 12-13
Economic exchange, 85-86
Economic growth: and The National Integration Program, 16
Economics: of ranching, 57
Ecosystems, 35
Ecuador, 7
Egypt, 156
EMBRAPA (Brazilian Enterprise for Agricultural and Cattle Ranching Research), 96, 234; and pasture recommendations, 55-56, 59; reserves of, 8; *see also* Research, agricultural
EMBRATER (Brazilian Enterprise for Technical Assistance and Rural Extension), 234; *see also* ACAR-PARÁ
Emigration, 116; of family units, 226; of individuals, 220-21, 226
Employment, 113, 148; *see also* Outside labor
Energy, 61-62, 83, 147; efficiency, 55
Enforcement of laws, 29
Enterprises, 26, 156; labor spent in, 210; of colonists, 113; ranching, 145; *see also* Corporations
Entrepeneurs, 111
Environmental degradation, 84, 118, 121, 157; carrying capacity assumptions concerning, 1
Environmental quality, 63, 125, 130, 156
Epidemiology, 123

Equilibrium: of populations, 61
Erosion, 40, 58, 102, 119, 122, 132, 147; modeling of, 199-200; under pasture, 56; under shifting cultivation, 55
Ethiopia, 51
Eucalyptus: deglupta, 32; *camaldulensis*, 33
Evaporation: generation of, 163-64
Evapotranspiration, 48
Experimentation: land use allocation for, 110
Experiment stations, 123
Exponential, 234; growth, 63-64, 68, 155, 157; trends, 6
Export: and shifting cultivation, 55
Extinction, 50-52; rates, 71

Faechem, R., 76
Failure: buffers against, 81, 125, 132; combined probability of, 139; criteria for, 79, 135, 143; criteria combined, 140-41, 144; definition of, 233; density independent, 81; margin for protection against, 172-74; maximum acceptable probability of, 79-81, 121-22, 125, 132, 136-39, 144, 173, 236
Failure probability/population density profile, 79-80
Falesi, I. C., 38-39, 55-56, 95, 227n2.1
Fallow, 62; effects on soil, 53-55, 59
Fallow periods, 18, 24, 123, 139-41, 171-72; in AGRISIM, 125; simulated (in KPROG2), 127; synchronicity of, 138-39
Family size, 122, 130, 215
FAO (Food and Agriculture Organization of the United Nations), 33, 114
Farinha, 234; charge for equipment used in making, 213; *see also* Manioc
Farmed time, 98
Feedback loops, 90, 128; in demographic transitions, 67
Felling, 26; cash requirements for, 183; financing of, 165-66; labor requirements of, 177-79; month distribution, 186
Fertility: human, 67, 69; 224-25; *see also* Demography
Fertilizers, 32, 123, 132; and limiting factors, 69; and nitrous oxide, 47-48; costs of, 184; effects on soils, 59, 195-99; in pasture, 56-57; prices of, 198; response to, 109, 154
Fiji, 15
Financing, 24, 100, 102, 124, 132, 165-70;

cash cost of obtaining, 169-70, 230nA.1; frequencies, amounts and terms, 166-67; payments of, 132; time spent in obtaining, 168-70, 230nA.2; *see also* Credit
Fiscal incentives, 28; *see also* Incentives
Fish: and várzea destruction, 149; purchases of, 124
Floods, 68
Florestas de rendimento, 33
Flow chart; of KPROG2, 131; *see also* Diagrams; Causal loop diagrams
Forcing functions, 89
Fordlândia, 19, 40
Foreigners: and motivation for the National Integration program, 17
Forest: exploitation, 33-34; reserves, 10-11
Forestry colonization, 33
Forestry policy law, 34
Forrester, J., 82-83, 85
Forrester-Meadows models; 234; *see also* Forrester; Meadows
Fortaleza, 117
Fossil fuels, 47, 62, 83, 86
French Guiana, 82
Frosts, 153
FUNAI (National Indian Foundation), 22, 234; reserves of, 8
Functional relationships, 90
Fungi, *see* Diseases
Fusarium solani f. *piperi*, 30; *see* Diseases (of crops), in black pepper

Gallery forest, 8
Game, 115, 124; price of, 213; sale of, 216; subsistence need for, 212; yield of, 211-12
Gastroenteritis, 224
Gaúchos, 97
Geopolitics: and the National Integration Program, 17
Georgia Pacific, 32
Germination, 123; in maize, 108, 201; in rice, 105; in *Vigna*, 203; of *Phaselous*, 202
Ghana, 109
Glassow, M. A., 77
Gleba, 234
Gmelina arborea, 32
Goiás, 37; deforestation in, 5; ranching in, 26
Goodland, R. J. A., 27, 52, 154, 227n7
Gourou, P., 75

Grasslands, 27
Greenhouse effect, 228n2.2; and deforestation, 43-45
Green revolution, 62
Grileiro, 23, 234-35
Growth: exponential, 63-64, 68, 155, 157
Guaporé Biological Reserve, 7
Guaraná, 31
Guatemala, 54
Guinea grass, 27, 57; *see also Panicum maximum*

Hardin, G., 150
Hayden, B., 77
Health, 98, 118, 132, 222-24; and lot abandonments, 118; effect on labor, 116; loss of work due to, 113
Hemileia vasatrix, 51
Hevea brasiliensis, 18; *see also* Rubber
Highways; routes of, 2
Holidays, 79
Homeostasis, 77
Hospital stays, 224
Houses, 20-24; financing purchases of, 166-68
Hubbell, S., 84
Humaitá, 36
Humus, 48
Hunting, 113, 124, 132, 209, 211-12; assignment of status for, 222
Hydrologic cycle: and deforestation, 48-51
Hylea, 37, 235

IBDF (Brazilian Institute of Forestry Development), 7-11, 93
ICOMI, 31
Ideologies, 151
Igapó, 8, 12, 35, 37, 149, 235
Illness: work days lost to, 224; probabilities by month, 223
Immigrants, 153
Immigration, 116; of family units, 226; of individuals, 226; probability of for individuals, 220-21
Incentives, 148; for pasture, 28, 151; tax and fiscal, 151
INCEPTISOL, 39
Inclusive fitness, 228n3.3
Income: nonagricultural, 125, 132, 209-11; *see also* Forest; Outside labor, distribution of; Wealth, distribution of

INCRA (National Institute for Colonization and Agrarian Reform), 19, 25, 93, 97, 154, 157, 235; colonization targets for Transamazon Highway, 227n6; credit from, 165-67; distribution of unsuitable rice seed by, 105, 118; yield expectation of, 172
Independent farmers, 111, 153
Indians, *see* Amerindians
Indigenous peoples, 52-53; *see also* Amerindians
Indonesia, 15
Inequalities; 147, 156; *see also* Wealth, distribution of; Income
Inflation, 6, 20, 28; adjustment for, 168, 230n5.3; and land speculation, 29
Infrastructure, 79
Injuries, 223-24; *see also* Trauma
INPA (National Institute for Research in the Amazon), 11, 235
INPE (National Institute for Space Research), 4, 235
Insects, 81, 102; *see also* Pests
Insolation: generation of, 164-65
Intensification: agricultural, 78
Intensive development, 151
Intercropping, *see* Interplanting
Interest: on loans, 125
International Bank for Reconstruction and Development, 151
Interplanting, 123; decisions, 174-75; in rice fields, 104-05
Investment: capital allocation to, 124; in lots and enterprises, 217-18; product allocation to, 113-14
Ipomea batatas, 216
IRHO (Institute de Recherche pour les Huilles Oleagenneaux), 31
Irwin, H., 227n7
Itaituba, 19; colonization area, 39

Jagunços, 23
Janzen, D. H., 41, 52, 62, 98, 155
Japanese-Brazilians, 19, 97
Jari, 32-33, 42, 236
Jaú National Park, 9
Java, 15
Jeito, 147, 236

K, 65
Kerosene: price and consumption of, 215

King Ranch, 27
Kleinpenninig, J.M.G., 16, 227n7
KPROG2, 85, 236; causal loop diagrams of, 128-29; documentation of, 121; features of, 121-25; flow chart of, 131; model structure, 126-30; modeling methods, 126; overdependence on pH, 139; parameters, 169; results of, 130-45

Labor, 24; affected by health problems, 116; calculations of, 130; checks, 132; constraints on allocations, 25, 125, 176-85; cost of, 176; equivalents by age and sex, 222; family, 113; for threshing rice, 219; hired, 113; patterns of, 124; price for felling, 219; requirements of agricultural operations, 177-82; requirements of male-only tasks, 176, 178-82; spent in outside occupations, 209-10; supply of, 124; types, 209-10; used in hunting, 209; *see also* Outside labor; Employment; Income, nonagricultural
Laborer farmers, 111, 141-44, 153
Land; distribution of, 7, 156; financing for purchase of, 166-69; *see also* Wealth, distribution of
Land clearing, *see* Clearing; Deforestation
Landless poor, 153
Land quality: classification, 85
LANDSAT, 3-4, 27, 236
Land tenure, 155
Land titles, 117
Land use allocation, 110-13, 127, 138, 165-86; assumptions concerning, 81
Land use: changing patterns of, 122; crops included in, 124; decisions, 100, probabilities for cash crops, 185; simulated, 142-43
Land use strategy: effect of colonist types on, 111
Land value, 28-29, 117, 151
Laws: circumvention of, 227n9; enforcement of, 29, 135-36, 147, 156
Laterite, 56, 58-59; *see also* Plinthite
Lateritic concretionary soil, 39
Leaching, 40, 57, 119, 147
Leaf cutter ants, 32
Learning, 125; in subsistence allocations, 172
Legal Amazon, 28, 35; changes in boundaries of, 227n1; definition of, 236; deforestation in, 4, 5; map of, 1-2
Lençois Maranhenses National Park, 9
Liana forest, 36
Licitação, 22, 30, 236
Lime, 123; price of, 198; soil changes from applying, 59, 195-96, 198-99
Limiting factors, 69, 82-84; multiple, 122
Limits to growth, 73, 229n3.7
Linear programming, 110
Linear response and plateau: model for crop yields, 201
Liquigas, 27
Living standard, 148
Loans, 28; as incentives, 148; payments of, 132; *see also* Credit, Financing
Lodging, 33
Logistic carrying capacity; *see* Carrying capacity, logistic
Logistic growth, 63-66, 70
Lots: abandonment of, 117; sale of, 117, sizes of, 138, 141
Lovejoy, T. E., 11
Ludwig, D. K., 32, 42

Macaroni: price, nutrition and consumption of; 215
Macroecological effects: of deforestation, 149
Magnesium: effects of agriculture on, 59; effects of burning on, 53; in soil under pasture, 55; analysis method, 102
Maintenance: of equipment, 219; of perennial crops and pasture, 132
Maize, 26; allocation to, 141-42; Amerindian land preparation for, 227n8; as cash crop, 112; cash requirements for planting, 183; disease in, 108; expected yield for, 172; fed to chickens, 124; financing of, 165-67; germination of, 108, 202; interplanted with *Phaseolus*, 175; interplanted in rice, 104-5, 174; interplanted in *Vigna*, 175; interplanting of, 175; labor requirements for, 177-79; planting density of, 201; price of, 213; rats in, 107-8; seed requirement for, 217; spoilage of, 211-12; subsistence need for, 214; technology change in, 109; yields of, 102, 123, 132, 200-1
Malaria, 51, 223-24
Malnutrition, 116
Malthus, T., 63, 65

Manacapuru, 31
Manaus, 65
Manganese, 31
Mangroves, 13, 37
Manioc, 26; allocation to, 141-42; Amerindian land preparation for, 227n8; bitter type in interplanted fields, 174; cash requirements for planting, 183; expected yield for, 172; growth period of bitter, 204; growth period of sweet, 204; in colonist diet, 115; interplanted in rice fields, 105; interplanted with maize, 108; interplanting of, 175; labor requirements for, 177-78, 180; price of, 213; subsistence need for, 214; sweet type in interplanted fields, 174; sweet vs. bitter type choice, 173; yields of, 102, 123, 132, 204-5; *see also* Root crops; Tubers
Manuring: effects on soil, 59
Marabá, 19; colonization area, 20, 39; ranch sales in, 21
Marajó, 36
Maranhão, 37; deforestation in, 5; ranching in, 26
Marasmius pernisciosus, 109-10; *see also Crinipellis perniciosa*; Diseases, in cacao
Marital status: effect of on emigration, 226
Marketing, 113, 147; difficulty of, 98, 125
Markets: commodity, 24, 31, 132, 148; weekly, 97
Markov matrices, 122
Marriage, 116; of colonists, 225-26
Marx, K., 151
Mato Grosso, 37; deforestation in, 4-5; land speculation in, 29; migration to Rondônia from, 23; pasture in, 26-27, 55-56
Maues, 31
Meadows, D. H., 73, 82, 85, 229n3.7
Meadows, D. L.; 229n3.7
Meat, 114-15; in colonist diet, 115; prices of, 213; priority placed on by Brazilians, 229n5.2; purchases of, 124
Mechanized agriculture, 94
Medicines: cash need for, 214
Medici, President Garrastazu
Meggers, B. J., 73, 228n3.3
Mesarovic, M., 116
Mesarovic-Pestle models, 236
Mexico, 156
Microcyclus ulei, 40; *see also* Diseases, in rubber

Migration: frequency of, 111; to Rondônia, 23; reasons for, 1
Milk nutrition: price/and consumption of, 215
Minerals, 16; and Amerindian reserves, 52
Minimum wage, 135-36
Mining, 31
Models: analytical, 86; simulation, 86-92; structure of, 91
Monetary correction, 168; *see also* Inflation, adjustment for
Monoculture, 40
Monsoon, 26
Morador, 97, 237
Moran, E. F., 18, 25, 95, 103, 111, 140, 152, 169, 219, 223-24, 227n7, 230nA.1, 230nA.2
Mortality, 223-24; of human population, 67; *see also* Death, rate; Demography
Murders, 118
Mychorrhizae, 40

National Integration Program, 16, 19-22
National Parks, 9-10
Natural regeneration forestry, 152
New Guinea, 15, 98
Newcomer colonists: initial capital and capital goods of, 222
Nicaragua, 85
Nietschmann, B. Q., 85
Nile River, 156
Nitrogen: change from virgin burns, 189-90; changes without burning, 193; changes under pasture, 194; effects of agriculture on, 59; in initial soil, 161-62; method, 102; in soil under pasture, 55
Nitrous Oxide, 46-48
Northeastern Brazil, 16, 18, 85, 117, 156-57
Northern Perimeter Highway, 8, 237
Nutrient cycles, 40, 62
Nutrients: and crop yields, 139; export of, 147; in virgin forest soils, 122
Nutrition, 133; and human fertility, 67; calculation of, 132; effect on fertility, 116; effect on mortality, 224-25; from goods bought with cash, 215; of colonists, 114-15; supplement from cash purchases, 216; supplement from tubes; value of food items, 115

Objectives: conflicts among, 150-52; of colonization programs, 153; of development, 147, 155
Odum, E. P., 84, 152
Odum, H. T., 62, 83, 85, 87
Oil (cooking): price and consumption of, 215
Oil palm, 31
Opportunity cost, 149, 151
Original colonists, 116
Outputs: of standard runs, 133-40; under alternative assumptions, 140-41
Outside labor, 125; distribution of, 155; land use strategy, 110-12
Overgrazing, 27
Overpopulation, 156; as a colonization motive, 7
Oxygen: and deforestation, 42
Ozone depletion, 46-48

Pacaás Novas National Park, 9
PAD (Directed Settlement Projects), 19, 237
Pandolfo, C., 33
Panicum maximum: production of, 27, 57, 205-6; *see also* Guinea grass
Pantanal Matogrossense National Park, 9
Papua New Guinea, 15
Pará, 36-37; deforestation in, 5
Paragominas, 27, 55-56
Parameters, 89; in statistics and modeling, 229n5.8
Paraná, 118
Parasites, 69
Pasture, 24-25, 145; allocation to, 141-42; and nitrous oxide, 48; as cash crop, 112; cash costs of, 184; burning of, 61; "carrying capacity," 72; degraded, 27, 154; effects on soil, 59; erosion under, 200; feeding capacity, 72; financing for, 165-67; incentives for, 28, 151; interplanted in rice fields, 105; interplanted with rice, 174; lack of fertilization in, 199; labor requirements for, 177-78, 181-82; opportunity cost of, 149; overgrazing of, 27; soil changes under, 45, 55-57, 132; 194-95; subsidies for, 148; technology change in, 109; yield of, 102, 123, 205-7
Patches, 113; size of, 126, 133
Perennial crops, 24, 30-31: land use strategy, 111-12; modeling yields of, 107, 109-110; technology change in, 109
Peru, 7, 57-58

Pesticides, 62
Pestle, E., 116
Pests, 51, 62, 81, 102, 123, 147, 154; reduction of under fallow, 54
Pepper, *see* Black pepper
pH: and aluminium toxicity, 98; and iron deficiency, 98; changes from liming, 196-98; changes from second growth burns, 190; changes from virgin burns, 188-89; changes from weed burns, 191; changes under pasture, 195; changes without burning, 192; effects of agriculture on, 59; effect of burning on, 53, 98; in initial soil, 161-62; in Transamazon soils, 38; map of, 96; method of analysis, 102; Overdependence on, 135
Phaseolus, 237; allocation to, 141-42; as cash crop, 112; disease in, 41, 107; effect of drought on, 41; expected yield for, 172; financing of, 165-67; interplanting in, 174-75; interplanting of, 202; planting density of, 201-2; price of, 213; probability of planting, 173; seed requirement, 217; spoilage of, 211-12; subsistence need for, 214; technology change in, 109; yields of, 102, 123, 132; *see also* Beans
Pharmaceuticals; from rainforest, 51
Phosphorus, 54; changes from fertilization, 198-99; changes from second growth burns, 190; changes from virgin burns, 189; changes from weed burns, 191-92; changes under pasture, 194; changes without burning, 192-93; effects of agriculture on, 59; effects of burning on, 53; effect on pasture yield, 205-7; effect on rice yields, 106; fixation of, 57; in initial soil, 163; in soil, 27; in soil under pasture, 56; in Transamazon soils, 38; method of analysis, 102
Photosynthesis, 61
Phytopthora palmivora, 109; *see also* Blackpod disease
PIC (Integrated Colonization Projects), 19, 237
Pico da Neblina National Park, 10
Pinus caribaea, 14, 32
Piper nigrum, *see* Black pepper
Pistoleiros, 23
Piúm, 118
Planting, 26
Planting dates: for rice, 105

Planting density, 123; of maize, 201; of *Phaseolus*, 201-2; of rice, 105-6
Plinthite, 56, 58-59
Plywood, 31
Podzolic soil, 57; *see also* ULTISOL
Political freedom, 17
Politics: and population policy, 68; and the National Integration Program, 16
Pollution, 149
Population: initial characteristics 116, 218; limiting behaviors, 228n3.3; pressure, 147, 149, 157
Population densities: and failure probability, 79-80, 144; effects on human fertility, 67; processes, 218-26
Population growth, 69
Population planning, 77
Population sector; 102, 116-19, 127
Potassium: effects of agriculture on, 59; in soil under pasture, 55; method of analysis, 102
Poverty: as a colonization motive, 7
Power saws; credit for, 165-67; depreciation of, 218-19
Prance, G. T., 50
Predators, 69
Predicted yields, 106
Prices, 132, 147; of goods bought with cash, 215; of land, 151; *see also* Land, value of; of products, 213; shadow, 151; variability in, 85
Primary productivity, 61
PROBOR, 30
Procrustean bed, 229n3.6
PRODEPEF, 34
Product allocation: modeling of, 209-18
Product allocation sector, 113-15, 127
Profit: influence on land use allocation, 110
Protein, 82, 130, 135; animal, 124, 228n3.4, 229n5.2; and population density, 137; consumed by colonists, 114-15; deficiency, 225; effect on mortality, 116; from fish, 149; from hunting, 222; requirements by age class, 22
Protein (total): supplement from tubers, 216
Pulp, 31
Purchases: in colonist diet, 115; of food, 114-15

Questionnaires, 100

Rabbits: attack in *Vigna*, 203
RADAMBRASIL, 3, 39, 58, 93-94, 144, 238
Rainfall, 41; and erosion, 122; Effect of deforestation, 44; 48-50; generation of, 163-164; in Altamira, 99; variability in, 99; of study area, 95
Rainforest: definition of, 238; protection, 7-14
Rainy season, 95; rainfall in, 164
Ranchers, 23; and Amerindian reserves, 52; and land conflicts, 153
Ranches, 21, 26; incentives for, 28; land use strategy, 111-12; subsidies for fertilization on, 57
Rangeland management, 72
Rats: damage to maize by, 107, 201
Reforestation, 33
Regeneration, 119
REGOSOL, 39
Regulations: circumventing of, 147
Reis, A. C. F., 17
Reis, M. S., 34
Relief, 93; *see* Topography; Slope
Remote sensing, 3
Research: agricultural, 95; reserves, 8, 11-12; value of rainforest for, 51; *see also* EMBRAPA
Reserves: Amerindian, 52; biological, 84; demarcation of, 155; Forest, 7-14; protection of, 149
Resource: allocation; distribution, 147; over-exploitation rate, 77; *see also* Land use allocation
Resources, 155; concentration of, 155; renewable, 149-51; *see also* Wealth, distribution of
Restinga, 37
Results; desviations from expected, 138-140; *see also* Outputs
Rhizobium, 40
Rhizoctonia: disease in *Phaseolus*, 202
Rhizoctonia microslerotia, 41
Rice, 26; allocation to, 141; as cash crop, 112; as dominant cash crop, 24; cash requirements for planting, 183; financing of, 165-67; expected yield for, 172; labor requirements for, 177-79; in colonist diet, 115; interplanting in, 174; interplanting with maize, 108; price of, 213; seed re-

INDEX 291

quirement, 217; spoilage of, 211-12; subsistence need for, 214; modeling yields of, 103-7, planting density, 104; technology change in, 109; varieties; 104-5; yields of, 102, 123
Rio Grande do Sul; 97, 234
Rio Negro, 35-36, 40
Risk: contracts, 34; maximum acceptable, 125
Roads: lateral, 20
Rondônia, 36; Deforestation in, 4-5; immigration to, 156; migration to, 23; population increase in, 6, 65; ranch sales in, 21; rubber germplasm in, 51; soil fertility in, 97
Root crops, 133
Root mat, 48
Root zone, 50
Roraima, 28, 36; deforestation in, 5
Rubber, 18-19, 30, 40; disease tolerance in, 51
Rurópolis, 238

Salati, E., 49
Sale of lots, 117
Salt: price and consumption of, 215
Sampling, 100-1
São Paulo, 17
Savana: soil compaction in, 54
Satellite imagery, 3; see also LANDSAT
Saturation density, 65, 238; see Carrying capacity, logistic
Sawlogs, 31
Schubart, H. O. R., 57
Scientific method, 88-89
Seasonality, 41-42, 69; see also Wet season, Dry season
Secondary forest, see Second growth
Second growth, 25, 48, 133, 140, 154; burning of, 61; cutting and burning dates, 187; effects on soils, 59; erosion under, 200; soil changes from burning, 190-91
Sectors: of KPROG2, 126-30
Seeds: determination of needs, 132; financing of, 165-69; government supply of, 124; improved varieties of, 122, 133; limit to allocations, 125; purchases of, 113; requirements of crops, 217; spoilage of, 211-12; varieties of, 86; see also Varieties
Selection of colonialists, 153

Self-sufficiency, 148, 153
SEMA (Special Secretariat of the Environment), 7, 12-13
Sensitivity testing, 89, 91, 126
Serrão, E. A. S., 56
Settlement, patterns, 18; see also colonization
Sex composition, 132; effect on labor, 116
Shifting cultivation, 18, 140; carrying capacity formulas, 74-76, 85, 238; carrying capacity studies of, 72-76; effects on soil, 53; see also Carrying capacity, for shifting cultivation
Shops, 94; credit from, 165-67, 217
Sidney Girão: colonization project, 23
Sigmoid curve, 63-64
Silviculture, 31-33
Simulation models, 86-92
Simulium amazonicum, 118
Sioli, H., 40, 42
Slash-and-burn agriculture, 79; effect on soil, 53; see also Shifting cultivation
Slope, 58; and erosion, 122, 199-200; generation of, 163
Smith, A., 151
Smith, N. J. H., 95, 211-12, 222-24, 227n7
Soap: price and consumption of, 215
Social injustices, 17
Soil: analysis methods, 102; average nutrient levels in, 130; changes from burning, 188-92; changes in, 123, 132; changes under pasture, 194-95; changes without burning, 192-94; compaction, 27; data collection, 102; degradation of, 81; density, 45; effect of deforestation on, 53-59; effect of erosion on fertility, 58; initial aluminum, 161; initial carbon, 163; initial clay, 162, initial nitrogen, 161-62, initial pH, 161-62; initial phosphorus, 163, initial quality of, 122, 132, modeling changes in, 186-200; on the Transamazon Highway, 38-39; under pasture, 132; variability of, 94
Soil erosion, see Erosion
Soil quality: in virgin forest, 85, 101-2; simulated changes in, 128
Solimões River, 28, 35
South American Leaf Blight, 40
Southern Brazil, 153
Soya Beans, 153
Species: diversity, 40-41; of trees, 93

Speculation: in land, 28, 52
Speculators, 23, 28-30; and land conflict, 153
Spoilage, 173-74; of products, 113, 123, 211-12
Squatters, 7, 12, 20, 23, 153; and deforestation, 156
Standard of living, 130, 135-36; carrying capacity assumptions concerning, 1
Standards: for carrying capacity, 84-85; see Carrying capacity
State variables, 89
Stochastic: definition of, 238
Stochastic processes, 128
Stochastic runs, 126, 136-37, 145
Storage: of food, 84
Street, J., 76, 80-81
Structure: of KPROG2, 126-30
Subsidies, 148; for fertilizers for large ranches, 57
Subsidy: from debt default, 216, through low-interest loans, 168
Subsistence: determination of needs, 132; need for cash, 214; need for products, 214; product allocation to, 113-14
Subsistence crops, 110; allocation to, 125, 141; used as cash crops, 185-86
Subsistence demand, 172
Substitutability, 83
Succession, 62
SUDAM (Superintendency for Development of the Amazon), 29; Deforestation estimates, 3; Forestry colonization plan, 33; incentives for guaraná, 31; incentives for pasture, 28; incentives for rubber, 30; income forest plan, 34; reserves, 14
SUDHEVEA, 30
Sugar: price nutrition and consumption of, 215
Sugar cane, 153
Suiá-Missu Ranch, 55
Sustainability: agronomic, 147; social, 147
System dynamics, 83
Swamp forest, 35; see also Igapó
Sweet manioc: as cash crop, 112; see also Manioc
Sweet potatoes, 216
Swift, 27

Tannia, 216
Tapajós River, 19, 36, 93

Taro, 79; see also Root crops; Tubers
Taxes: and deforestation motivations, 154
Tax incentives, 28, 148; see also Incentives
Technical assistance, 16
Technology: assumptions concerning, 1, 72, 81; changes in, 86, 121-122; factors for change in, 109; improvement of, 107; simulated change, 132
Tefé, 31
Teitelbaum, M., 68
Temperature: global change from deforestation, 43-45
Terra firme, 35-36, 97; area of, 3
Terra preta do Indio, 38, 238
Terra roxa, 57, 94-95, 97, 238; decline in estimated areas of, 227n2.1; on Transamazon Highway, 38-39
Thanatephorus cucumeris, 41
Threshing machines: depreciation of, 218-19
Timber: destruction of by shifting cultivation, 55
Time horizon, 133
Titles: for land, 29, 117
Tocantins River, 8
Topography, 93, 101, 163; see also Slope
Toppling: in rice, 104-5, 123
Torture, 17
Tragedy of the commons, 150
Transport: costs of, 230nA.1; to markets, 212-13
Transportation, 98, 101, 132, 154; availability of, 125; cash need for, 214; cost of, 169; difficulty of, 113; probability of availability, 213
Transamazon Highway; 1-2; bisection of park by, 9; building of, 16; colonization projects on, 19; colonization targets, 20; description of, 238; pastures on, 27; population absorption by, 157; soils on, 38-39
Transition forest, 28
Transmigration, 15
Trauma, 223-24
Travessão, 239
Tribal peoples, 52-53; see Amerindians
Trinidad, 107
Trombetas River, 36
Tubers, 133; in colonist diets, 115; nutritional supplement from, 216
Tucuruí, 8
Tumucumaque, 8

INDEX 293

Turnover of colonists, 141; effect on capital and capital goods, 222
Turtles, 85

ULTISOL, 39, 57-58, 94, 155
Ultraviolet radiation, 46-48
Underclearing, 26, 165; see also Broca
Understory, 26
UNESCO (United National Educational and Scientific Organization), 15, 229n3.5; 234; nutrition standards, 229n5.2
Upper Amazon, 28; see also Solimões River
Urban centers, 121, 155

Validation, 89; 91
Variable: in statistics and modeling, 229n3.8
Variability: effect on carrying capacity, 81-82, 88; effect on failure probability, 136 and failure probabilities, 144-45; in market prices, 85; of soils, 94; in the colonist population, 101; in yields, 125; in weather, 122; modeling of, 107; of soil pH, 96
Varieties: improved, 122; of rice, 105
Várzea, 28, 35-37, 97, 149, 154, 239
Vegetables, 19
Vegetation types, 35-37, 93
Venezuela, 7, 98, 227
Verhulst, P. F., 63, 70
Vertebrate pests, 102
VERTISOL, 39
Vigna, 239; allocation to, 141-142; as cash crop, 112; definition of, 239; expected yield for, 172; price of, 213; probability of planting, 173; seed requirement, 217; spoilage of, 211-12; subsistence need for, 214; technology change, 109; yield of, 102, 123, 132, 203-4
Vines, 26; see also Liana forest
Violence: in pioneer areas, 97

Virgin forest, 26; definition of, 229n4.3; burning in, 53
Volkswagen ranch, 27

Wage labor, 132; see also Outside labor
Water stress, 50
Wealth: distribution of, 1, 68, 147, 149, 155
Weather, 68, 102, 128; generation of, 163-65; simulation of, 122, 132; variability in, 165
Weeding, 26, 61
Weeds, 81, 138; erosion under, 200; and fallow periods, 54; in pasture, 56; limit on cultivation, 171; soil changes from burning, 191-92
Wet season, 25; see also Rainy season, 95
White sand, 36
White water rivers, 35
Wife: probability of presence and age of, 218
Witches' broom disease, 109-10
Women: outside labor of, 210
Wood volume, 93
World 3, 229n3.7
World Bank, 151
World Models, 85, 116, 82; see also: Meadows, D. H.; Forrester, J.; Mesarovic
WWF-US (World Wildlife Fund-US), 11

Xanthosoma, 216
Xingú River, 8, 36, 93

Yams, 216
Yield: expected, 172
Yurimaguas, 58

Zambia, 82
Zona Bragantina, 4, 18-19, 239; definition of, 239; liming experiments in, 196, 198

Please remember that this is a library book, and that it belongs only temporarily to each person who uses it. Be considerate. Do not write in this, or any, library book.

DATE DUE